Electric Motors and Their Applications

Electric Motors
and Their Applications

TOM C. LLOYD
formerly Chief Electrical Engineer
Vice-President, Electrical Engineering
Robbins & Myers, Inc., Springfield, Ohio

WILEY-INTERSCIENCE
A DIVISION OF JOHN WILEY & SONS
New York · London · Sydney · Toronto

To The Memory of Three Unsung Martyrs

Thomas Davenport of Brandon, Vermont, who invented the electric motor (U. S. Patent #132, 1837) and died penniless as a result.

Emily Davenport, his wife, who gave up her silk wedding dress for insulation on the windings of the first electric motor.

Oliver Davenport, itinerant merchant, who sold his horse and wagon to finance his brother's experiments, thereby putting himself out of business.

10 9 8 7 6 5 4 3 2 1

Library of Congress Catalog Card Number: 70-77834

SBN 471 54235 0

Printed in the United States of America

Preface

In this affluent society the average American home may contain about 20 electric motors. To every member of the family they are just electric motors, although they obviously differ in size and mechanical design. Perhaps the discerning householder may have discovered that a few appear to spark inside, others make a clicking noise when started, and others do neither.

The differences go much deeper than that. These 20 motors probably represent six different principles of operation.

The number of types is further increased when one sets foot in an office building or a factory and examines the variety of motors in use there.

The operational characteristics of each can be understood with simple mathematics and some fundamental electrical theory. These characteristics form the reason for using one type of motor on the dishwasher, another on the vacuum cleaner, and still another on the electric typewriter.

A few of these devices could operate satisfactorily with a different and cheaper type of motor. A few could operate with motors that are larger or smaller, with the same horsepower output, and bought at a reduced price.

This is a challenging statement.

It is based on the belief that some engineers, who design the vast amount of motor-driven equipment found on the market, do not know enough about the motors that are available.

This situation is brought about also by the way in which many motor-driven devices apparently are designed. The process involves these steps:

The designer develops a new domestic ironer, desk calculator, dishwasher, or what not. He uses great ingenuity in producing a sample that fulfills its primary function and can be produced economically.

Little space remains inside the device for the motor.

The sample, in its prototype cabinet, is then given to the industrial designer, who smooths out the corners, "reduces the silhouette," and gives the product eye appeal.

Now there is still less space inside for the electric motor.

He then calls in the motor application engineer (salesman) and says that he needs a production motor to drive his device.

This procedure can result in the need for special motors of non-standard mechanical sizes, special motors in which the losses must be minimized because of overheating, and in general a more expensive product than might have been required had the motor supplier been called in at the inception of the development.

The above account is exaggerated, of course, but it happens occasionally to the full degree as described and more often to a lesser degree.

A more widespread knowledge of motor types, operation, and standards might result in reduced costs.

It is hoped that this book, which is the outgrowth of a number of courses and talks I have given over many years, will help to bring this about. My audiences consisted of salesmen, factory superintendents and foremen, technicians, and laboratory assistants. All of these people had been involved in various phases of production, sales, and testing of electric motors.

As an engineering executive, however, I had a natural interest in helping the sales department to supply engineering information on electric motors to prospects and customers. Some of the chapters in this book therefore are based on almost identical information supplied to purchasing agents and engineers of various backgrounds who were in a position to buy and/or specify electric motors for their products. They became curious about how motors work and about the relative merits of the various types.

Three generations of electrical engineers have burned midnight oil to develop the theory and mathematics for analyzing and designing electric motors. Today's designers are working equally hard with computers in an attempt to design motors to meet the almost infinite variety of performances sought by motor users.

I owe them all an apology for minimizing their mathematical wizardry and attempting to make their achievements appear simple.

Acknowledgement is due to many former associates at Robbins &

Myers, Inc., who helped to supply data for this book or who sat patiently through the lectures that formed the material used. Special mention should be made of John A. Zimmerman, Manager of Marketing, who for many years goaded me into writing it.

Yellow Springs, Ohio   T. C. LLOYD

Contents

Electric Motors and Their Applications

Chapter 1

Electrical Fundamentals, Electric and Magnetic Circuits, Generator and Motor Action

To understand electric motors it is necessary to have a knowledge of the terms used and the nature of the electric and magnetic circuits. These terms are introduced by considering direct-current (dc) circuits and the magnetic circuit as set up by direct currents.

Interaction of the electric and magnetic circuits leads to a basic understanding of motor and generator action.

The concepts of direct current are useful not only for their own sake (for dc equipment has seen a revival in recent years) but also because they provide an introduction to the more complex alternating-current (ac) circuits.

1-1 The DC Circuit

If we connect a long loop of wire to the terminals of a battery, an electric current will flow through the wire. It is forced through the circuit by the electric pressure which is measured in *volts* (v). This pressure is also called *electromotive force* (emf). The rate of current flow is measured in *amperes* (amp). The current is limited by the *resistance* of the wire and this resistance is measured in *ohms* (Ω).

The resistance to current flow is determined by the material of which the wire is made, its length, and its cross-sectional area. To a lesser extent it also depends on the temperature of the wire. Thus, if we double

1

the length of the wire, one-half as much current would flow. In general we can represent many natural phenomena by this expression:

$$\text{effect} = \frac{\text{effort}}{\text{opposition}}. \qquad (1\text{-}1)$$

The effect we obtain in this electric circuit is a flow of current. The effort is through the pressure or voltage built up by the battery and the opposition is the resistance of the wire.

To apply this general expression specifically,

$$\text{amperes} = \frac{\text{volts}}{\text{ohm}}. \qquad (1\text{-}2)$$

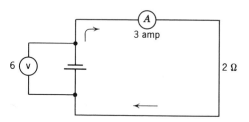

Figure 1-1. A loop of wire is connected to a 6-v battery. The resistance of the wire is 2 Ω. By Ohm's law we can calculate that the current in the wire would be 3 amp as measured by the ammeter, in a series with the line. Voltmeters are connected *across* various points in a circuit as they measure the difference in electrical potential between those points.

Using the common symbols, I for amperes, and R for ohms, we have

$$I = \frac{V}{R}. \qquad (1\text{-}3)$$

This is known as *Ohm's law*. Our forefathers were kind enough to define these units so that we can say that 1-v causes 1 amp to flow against 1 Ω resistance without introducing other numbers as constants (See Figure 1-1).

Ohm's law applies not only to an entire circuit but also to the parts thereof. Suppose a concentrated resistance such as a light bulb of 1 Ω is connected to the battery through lead wires each of $\frac{1}{2}$-Ω resistance (see Figure 1-2). Now the total resistance of the circuit is

$$1 + \tfrac{1}{2} + \tfrac{1}{2} = 2 \text{ ohms.}$$

The current is then,

$$I = \frac{V}{R} = \frac{6}{2} \text{ or 3 amp.}$$

Since $I = V/R$, it follows by simple algebra that

$$V = IR, \tag{1-4}$$

that is, the voltage across a part of a circuit will equal the product of amperes and ohms (refer again to Figure 1-2). If a voltmeter were connected from a to b, it would read

$$3 \times \tfrac{1}{2} = 1\tfrac{1}{2} \text{ v.}$$

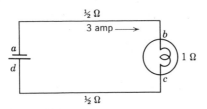

Figure 1-2. A lamp of 1-Ω resistance is connected to a battery by leads, each having a resistance of $\tfrac{1}{2}$ Ω.

Connected from c to d it would read the same. Connected from b to c it would read

$$3 \times 1 = 3 \text{ v.}$$

We have applied 6 v to this circuit. Of this, $1\tfrac{1}{2}$ v are "used up" in overcoming the resistance of one lead wire, 3 v are applied to the lamp, and the other lead wire "uses up" the remaining $1\tfrac{1}{2}$ v. Thus, of the 6 v applied, all are represented by an equal number of "voltage drops" around the circuit. This is always the case. It is an example of action and reaction being equal and opposite.

If we used a 12-v battery on this circuit, the current would double and the "voltage drops" would double, equaling the 12 vs applied. Note that the current, being the *effect* in this phenomenon, adjusts itself to bring about this balance.

Note also that, although we applied 6 v to this circuit, only 3 v appeared at the lamp terminals. This is an exaggerated case of what might happen when a large electric motor is connected to the end of a long line in which the wire cross section is too small and, therefore, has a high resistance. The motor would not get the benefit of full line voltage.

1-2   Power in the DC Circuit

Power is the rate at which work is done. In the electric circuit this is measured in *watts* (w) and it ties into horsepower (hp) by ratio of 746 w being equal to 1 hp. In dc calculations, watts equal the product of current and voltage.

It must be emphasized that power is a *rate*. Thus, if we consider a 100-w lamp or a 400-w motor, these terms measure the rate at which power is being used. The total *energy* used involves *time* and power, measured as kilowatthours (kwhr), which is what the electric user pays for, as it represents the total amount of work done, electrically, in a given period.

Refer to Figure 1-2. Here we had a voltage drop of $1\frac{1}{2}$ v in each wire when 3 amps were flowing.

$$\text{watts in each} = 1\tfrac{1}{2} \times 3 = 4.5 \text{ w.}$$

In the lamp itself

$$\text{watts} = 3 \times 3 = 9 \text{ w.}$$

Total watts of circuit:

$$4.5 \text{ (one wire)} + 9 \text{ in lamp} + 4.5 \text{ (other wire)} = 18 \text{ w.}$$

This can also be calculated from the battery output. It supplied 3 amps at 6-v pressure.

$$3 \times 6 = 18 \text{ w.}$$

Hence the battery puts out energy at exactly the same rate at which it was being used in the circuit.

Since power $= V \times I$ in watts and

$$V = I \times R,$$

it follows that

$$\begin{aligned} \text{power} &= (I \times R) \times I \\ &= I^2 R. \end{aligned} \tag{1-5}$$

Consider again the lamp in Figure 1-2. Its resistance is 1 Ω. The current was 3 amps. The watts are then $I^2 R = 9 \times 1$ or 9 w. This is verified by the previous calculation.

The term "$I^2 R$ loss" is frequently used in dealing with circuits and motors.

1-3   Parallel Circuits

Figure 1-2 showed what were in effect three resistances in series. The voltages varied across parts of the circuit, the current was the same throughout the entire circuit. Now refer to Figure 1-3. Here we have two resistances connected across a 100-v line in parallel. The resistance of the leads will be ignored.

What is the current in each part of this circuit?

$$I = \frac{V}{R}$$

$$= \tfrac{100}{20} = 5 \text{ amp} \qquad \text{in part } (a)$$

$$= \tfrac{100}{10} = 10 \text{ amp} \qquad \text{in part } (b)$$

$$= 5 + 10 \text{ or } 15 \text{ amp} \qquad \text{in part } (c).$$

Note that the combined effect of two resistances in parallel shows less resistance than that of any one branch.

$$R_T = \frac{V}{I} = \frac{100}{15} \text{ or } 6.66 \ \Omega.$$

Resistances in parallel can be added by obtaining the sum of the reciprocals. (This sounds more complicated than it is.)

$$\frac{1}{R_T} = \frac{1}{Ra} + \frac{1}{Rb,} \qquad\qquad (1\text{-}6)$$

$$\frac{1}{R_T} = \frac{1}{20} + \frac{1}{10} \text{ or } 0.05 + 0.10 \ \Omega,$$

$$\frac{1}{R_T} = 0.15,$$

$$R_T = \frac{1}{0.15} \text{ or } 6.66 \ \Omega.$$

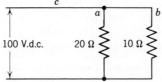

Figure 1-3. The conventional method of representing resistance employs the sawtooth lines as shown. These two resistors are in parallel.

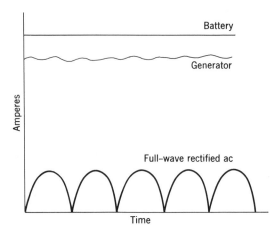

Figure 1-4. These currents are all defined as dc because they represent flow in one direction. Special considerations must be made in dealing with the lower, rectified quantity.

This checks with the resistance obtained by using the total current in the circuit (see Figure 1-4).

1-4   The Magnetic Circuit

We are all familiar with permanent magnets, but let us consider one shaped like a straight bar, as shown in Figure 1-5. Placed under a sheet of glass on which iron filings have been placed, we note some form of magnetic force which lines up the filings into the paths as shown. This magnetic field is considered as being made up of magnetic lines of force or flux lines. They are considered as coming out of one end, called the north pole, and entering the other, called the south pole. Each line must make a complete circuit, through the magnet and loop around from one end to the other. There is no motion along them; in fact, the flux line is entirely an imaginary concept. But it is inadequate in engineering work just to say that this is a strong field or a weak field. Quantitative measurements are needed. With the concept of lines we can then measure a field strength by saying it has 40,000 lines per square inch (in.^{2}), or 100,000 lines per in.^{2}. The kilogauss[1] is also a common unit, being 1,000 lines per square centimeter (cm^2). Although the "line"

---

[1] Named for a German mathematician, Gauss. One of his lesser known accomplishments was the development of a formula by which the date of Easter could be calculated.

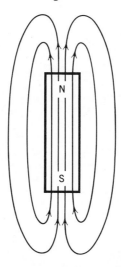

Figure 1-5. Magnetic field built up by a permanent magnet.

is imaginary, in dealing with it consistently in all related phenomena it is mathematically useful and correct. The symbol B is used to denote field intensity.

1-5   Magnetic Fields and the Electric Current

Whenever a current flows in a conductor, a magnetic field is built up around it. This is a natural phenomenon. Refer to Figure 1-6. Here a

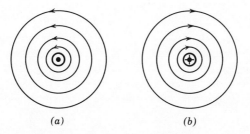

(a)   (b)

Figure 1-6. If we imagine that a conductor is clasped in the palm of the right hand, with the current flowing in the direction of the thumb, the flux is built up in the direction of the fingers. (a) Cross section of a conductor with the current flowing out of the page. This is represented by the dot, indicating the head of an arrow. (b) Cross section of a conductor with the current flowing into the page. The "cross" is the tail of an arrow.

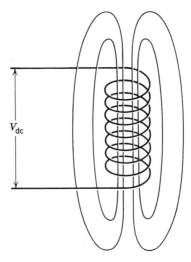

V_{dc}

Figure 1-7. Magnetic field built up around a coil with current flowing.

current is shown flowing first in one direction and then reversed. The magnetic field has a definite direction with respect to the current as shown. If the current goes to zero, the field collapses to an imaginary line in the center of the conductor. Reversing the current reverses the field. These lines extend out indefinitely into space, but they fall off in intensity as the inverse square of the distance. That means that, if we have a field intensity of 200 lines/in.^{2} close to the conductor, at a distance of 10 times further out, the intensity will be $\frac{1}{100}$ or 2 lines/in.^{2}. This explains why amateurish "inventions" which depend upon magnets working through great distances are never successful.

If the conductor is wrapped in the form of a coil, as shown in Figure 1-7, the fields about each turn add up to a flux system similar to that displayed by the permanent magnet. For convenience we consider that, as a direct current is made to flow through this coil, the flux builds up from zero from an imaginary line through the center of the coil to a position as shown. Opening the circuit causes the flux to collapse to this line.

1-6   Effect of Iron or Steel

If we would place a piece of steel inside the coil, the flux would be greatly increased. If the steel were a complete loop so that the entire

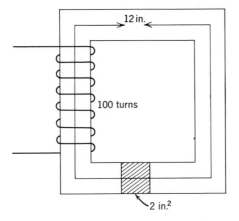

Figure 1-8. A steel core of laminations is built up to the dimensions shown and is magnetized by a coil of 100 turns.

magnetic circuit could be set up in steel, the flux would increase a great deal more. This construction is shown in Figure 1-8.

In addition, if we increase the current in the coil, more flux is built up, which also occurs if the number of turns is increased. The flux system around each added turn adds to the total. Examine the general rule:

$$\text{effect} = \frac{\text{effort}}{\text{opposition}}.$$

We can conclude that the effort expended, depending on both the number of turns and the amperes, is their product, N (turns) $\times I$. This is the magnetizing force. The effect is a flux system of ϕ lines with an intensity of B lines/in.^{2}.

By adding steel to the circuit and obtaining more flux lines thereby, we can see that it is easier to set up flux in steel than in air. The opposition to the setting up of flux is called the *reluctance*. The inverse of this, the ease with which flux is set up, is called the *permeability*.

We can now fill in our equation for the magnetic circuit:

$$\phi \text{ (flux)} = \frac{NI}{\text{reluctance}} K. \tag{1-7}$$

Magnetic circuits are cursed with many units of measurement, each requiring a different definition or value of K. We will deal only with the simplest method.

1-7   Variable Reluctance

Only iron, steel, and some materials alloyed with them are classed as *magnetic materials*. Air, paper, aluminum, wood, etc., all have the same reluctance. Unfortunately the reluctances of the magnetic materials are not constant. If we have only a small flux set up in the magnetic core, as shown in Figure 1-8, doubling NI may double the flux. But if we have many flux lines through this structure, doubling NI may add only 25 or 50% more flux. This is because iron and steel have the unfortunate property of saturating magnetically. For the electric circuit this would be comparable to saying that, if we already have 10 amp flowing in a line and double the voltage applied, we would not have 20 amp but somewhat less.

As a result, our approach to making calculations of magnetic circuits must depend upon having actual tests results of NI versus ϕ or B, pertaining to the material to be used. A typical curve is shown in Figure 1-9.

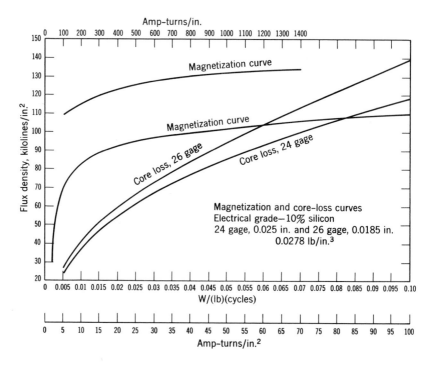

Figure 1-9. Magnetization curves for two gages of steel of the same composition.

1-8 Example

We wish to determine how many amperes are required in the coil of Figure 1-8 to set up a flux ϕ of 160,000 lines in the core. The core is made up of sheet steel and its cross section is 2 in.^{2}. The flux density in lines/in.^{2} is therefore 160,000 divided by the 2 in.^{2} or 80,000.

Refer to the curves of Figure 1-9. This flux density requires 8 amp-turns/in. of length from the magnetization curve for this sheet steel. With 12 in. of total path, the amp-turns required are 12×8 or 96. As there are 100 turns on the coil, the current required would be 0.96 amp.

This has all of the appearance of being a precise calculation. But if the steel used does not conform exactly to the curve as shown, naturally the calculations are inaccurate.

1-9 Effect of an Air Gap

Assume now that the magnetic core of the previous example has an air gap as shown in Figure 1-10. It is 0.015 in. long, and again we wish to set up 160,000 lines in the core and across the gap.

Again 8 amp-turns are required per inch of steel path in the core. This will still be considered as being practically 12 in., and so 96 amp-turns are required. But as to the gap itself, with the inch units we are using, a constant of 0.313 in. is required.

$$
\begin{aligned}
\text{amp-turns for air path} &= 0.313 \times B \times \text{length of gap} \\
&= 0.313 \times 80,000 \times 0.015 \text{ in.} \\
&= 375. \qquad\qquad (1\text{-}8)
\end{aligned}
$$

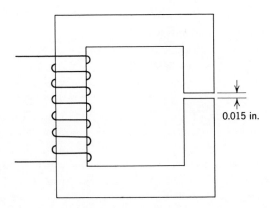

0.015 in.

Figure 1-10. The core of Figure 1-8 now has an air gap in its flux path.

We now need 96 plus 375 or 471 amp-turns to set up this flux through the steel and air gap. Note the proportion of magnetizing force needed between 12 in. of steel and 0.015 in. of air. It is obvious that air has much greater reluctance than steel. The current required in the coil would be 4.71 amp.

It would be of interest to see the effect of attempting to set up 50% more flux through the core. Now the density would be 120,000 lines/in.^{2}. Use the upper curve and top scale of Figure 1-9 and note that 300 amp-turns/in. are required. For the 12 in. (ignoring the short gap), 3,600 amp-turns would build up the required density.

For the gap

$$NI_{gap} = 0.313 \times 120,000 \times 0.015 \text{ in.}$$
$$= 562.$$

Total amp-turns required are 4,162 and the current would be 41.62 amp.

Much can be learned from the foregoing examples. Air does not saturate, and an increase of 50% in flux density requires a 50% increase in magnetizing current. The steel, however, shows the effect of saturation so that an increase of 50% in flux increased the amp-turns by 3600/96 or 37.5 times.

The magnetic circuit of this core with an air gap is, in essence, the same as the field of an electric motor. It can be expected that, if the steel in the motor is "worked too hard" magnetically, the motor will have to draw a very heavy current from the line to build up the flux expected. To obtain practical values of motor currents, the flux density must be kept reasonably low. This is true of both dc and ac motors.

1-10   Generator Action

Consider a coil with an iron core having an air gap as shown in Figure 1-11. Near this large air gap between the pole faces we shall place two conductors. The top conductor is going to be moved to the right through the flux of the gap. As a result, a voltage is generated which could be measured by a voltmeter connected to the ends of the conductor. Move it faster, and a higher voltage is generated. This represents a natural phenomenon, identified as *generator action*. Note that the flux partly wraps around the conductor, and it is from this that we can determine the direction in which the voltage is set up.

Note the lower conductor moving to the left. Here the flux partly

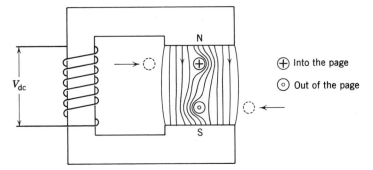

Figure 1-11. When a conductor is moved through a magnetic field a voltage is generated in it. With the fingers of the right hand in the direction of flux and the conductor moving into the palm, the thumb indicates the direction of the voltage generated.

wraps around the conductor in the opposite direction and the generated voltage is reversed.

We know also that cutting through 10^8 (100,000,000) lines of flux/sec generates 1 v in a conductor.

If the conductor is moved straight up and down along the flux, no voltage is built up. If the conductor is held stationary and the field structure is moved across it, a voltage is generated. Relative motion across the flux is the important factor.

1-11   Induced Voltage

We are now in a position to tie together what we know about the electromagnetic circuits and the basis of generator action. Consider the coil of Figure 1-12 which is going to be connected to a dc source. No magnetic field exists around the coil until the current flows. Note the extra loop at the bottom of the coil. At one instant there is no flux through this loop, an instant later, with current flowing in the main coil, the flux system extends through the loop. If the loop were open and a voltmeter were connected to its terminals, it would be noted that a voltage was generated only during that time in which the flux was building up from zero to its final steady value. Once the flux is constant, there is no cutting.

Close the loop during this time and a momentary current flows.

Cut off the current in the main coil. Now the flux collapses to zero,

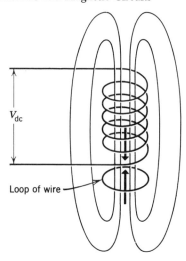

Figure 1-12. Adjacent to a coil connected to dc source is a single loop of wire. The arrows indicate the direction of flux set up when the supply circuit is closed.

cutting the loop in the opposite direction. A momentary voltage is again built up causing current to flow in the opposite direction.

If we would use the right-hand rule to determine the direction of flux set up in the coil for a given current flow, then use it also to determine the direction of voltage and current set up in the loop, we would find that the flux of the loop opposed that of the coil when the current is building up. Common sense would also tell us that this had to be the case. If the flux of the loop *added* to that of the coil, the increased flux would generate still more voltage in the loop. More current would flow, more flux would be built up, and so on *ad infinitum*.

We have explained the foregoing phenomenon on the basis of the flux cutting the loop and generating voltage. This is still true, but a more conventional approach in a case like this is to consider that voltage is induced in a coil owing to the flux *linking* it. A change in flux linkage induces a voltage. Again 10^8 flux linkages/sec. induce 1 v. Obviously, if our loop had 10 turns, the voltage of each turn would add, giving 10 times the voltage induced.

1-12   Self-Inductance

In the foregoing discussion it was pointed out that a changing flux through a separate loop or coil resulted in an induced voltage. But what about the coil itself to which the direct current was applied? Just before the switch was closed, there was no flux. An instant later a whole flux system is built up, coming out from this imaginary line in the center of the coil and cutting all of the turns in the process. Obviously, this must induce a voltage, and, by checking by the right-hand rule, it follows that the induced voltage in the coil opposes the applied voltage.

The greatest voltage is built up when the flux linkages are changing at the greatest rate. This occurs at the instant the switch is closed and limits the current in the coil. The current then builds up gradually, not instantaneously, and, as the *rate* at which it builds up is reduced, so also is the induced voltage (see Figure 1-13).

Unfortunately, in dealing with coils and voltages induced, we cannot consider that every flux line links every turn in the coil. This can vary through coil configuration, the presence of iron or steel, and its saturation. It is therefore necessary to define this phenomenon known as *self-inductance*. It is measured in *henrys* (h).[2] A henry is the inductance of a closed circuit which induces 1 v when the current varies uniformly at the rate of 1 amp/sec. It is abbreviated as L, or

$$L \text{ is proportional to } \frac{N\phi}{I},$$

$$L = \frac{N\phi}{I} K. \qquad (1\text{-}9)$$

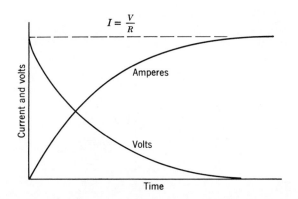

Figure 1-13. When the circuit is closed the initial growth of current and flux induces a voltage. Only the difference in the applied and induced voltage is available to force current to flow at that instant. When the current finally reaches its steady value the induced voltage becomes zero.

[2] Named for the American physicist Joseph Henry. Faraday in England noted the phenomenon of self-inductance about the same time. Another unit was named for him.

The constant K depends on which of an unfortunately large number of systems of units is used.

1-13   Motor Action

Let us return to the magnetic circuit with a dc excited field and a large air gap. Refer to Figure 1-14, in which it will be noted that a conductor, carrying current, is placed in the edge of the flux system of the gap. The conductor alone would build up a flux system because of its current, as already discussed.

When this conductor is placed in a magnetic field, there is a reaction between the two as shown. The "piling up" of flux behind the conductor exerts a force on it so that it moves across the gap until it leaves the boundary of the flux built up between the pole faces. If we reversed the current in the conductor, it would move back to the right. Or if we reversed the direct current in the coil, the conductor would move to the right also.

Here, again, is a simple natural phenomenon, known as *motor action*. Nearly all electric motors operate because of this simple fact. The force exerted on the conductor is proportional to the intensity of the flux between the pole faces and the current in the conductor. This has a great influence on motor design and the size of a motor for a given horsepower and speed.

Although we show a circular flux system around a conductor in air and arrows indicating the direction of the flux, there is actually no

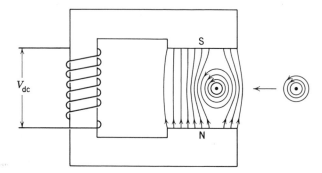

Figure 1-14. A conductor carrying current builds up a circular magnetic field. Placed in another field built up between the pole faces of an electromagnet, the reaction of the two fields exerts a force on the conductor, moving it across the field. This shows the elements of motor action.

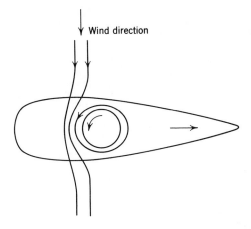

Wind direction

Figure 1-15. An experimental "sailing" vessel was built in which the "sail" was a large diameter stack rotated by an engine. The surface air rotated with it, reacting with the wind. Air pressure built up behind the stack exerting a forward force. Although there was actual motion of the air (unlike the flux lines), note the comparison with "motor action."

motion along these lines. This is also the case with the flux between the pole faces.

It is of interest to see how a similar system involving actual motion around the components illustrates this motor action.

About 50 years ago an experimental sailing vessel was built in Europe. The hull was of the usual type but the "sail" consisted of one tall stack of large diameter. The stack was rotated by a gasoline engine (see Figure 1-15). The rotating stack piled up air pressure behind it and created a partial vacuum at the front. As a result the vessel operated most satisfactorily sailing across the wind. The effectiveness of this construction is not known, although the fact that it was never duplicated probably indicates that it was not a great success.

Chapter 2

Resistance and Inductance in AC Circuits

2-1 The Elementary AC Generator

Figure 2-1 shows a loop of wire arranged to rotate in a magnetic field. The ends of this loop are connected to slip rings. These rotate with the loop, but there are stationary brushes which bear on these rings. Leads to the brushes enable electrical contact to be maintained between an exterior circuit and the rotating loop. The loop is an elementary armature winding.

As the loop rotates, it will cut the flux and generate a voltage. But we know that "sliding along" the flux does not result in cutting at the same rate as movement perpendicular to it. We can assume, therefore, that the voltage generated will vary with different positions of its rotation. To deal with this accurately, and we must, we shall digress a bit with some elementary trigonometry.

2-2 A Few Concepts from Trigonometry

Consider a line of 100 units in length rotating about the point 0 in a counterclockwise direction. As it rotates, we are interested in the vertical projection of this line (see Figure 2-2). By the time it has reached point 2, its height is 70.7 units. At point 3 its full length is vertical. At point 4 it is again 70.7 and continues to decrease until its vertical projection is zero at point 5. The values of the vertical projections are plotted as shown to form an upper loop of the curve. (Actually more intermediate points would be required to fix the curve more accurately.)

As the line continues to rotate, it projects downward, representing

18

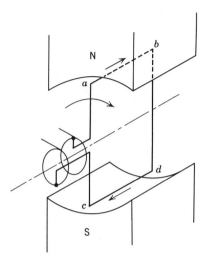

Figure 2-1. An elementary ac generator. The armature winding, shown as only one loop, rotates in a flux system set up by coils on a field structure.

negative values. The completed curve is called a *sine wave*. The distance a-b, generated by making one revolution, can represent either 360° or the time required to make the revolution.

The height under the curve at a definite point is the same as that of the triangle shown in Figure 2-3. For 45°, with a diagonal of 100 units, the height y is 70.7, and so on, for each angle turned through.

Instead of referring to the ratios of the sides to the hypotenuse as

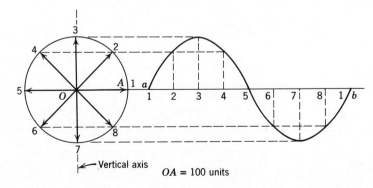

Figure 2-2. As the line OA rotates, its vertical projection varies in the manner shown by the curve.

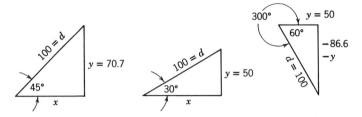

Figure 2-3. The ratio of the side y to d varies with the angle. For convenience y/d is called the *sine* of the angle. Similarly, x/d is called the *cosine* of the angle.

y/d or x/d, it is more convenient to use the trigonometric terms *sine* of the angle and *cosine* of the angle respectively. Condensed tables of these values are given in the Appendix.

2-3 Generated Voltage Waves

Refer to Figure 2-4, which is the same generator shown in Figure 2-1 without the completed coil and slip rings and brushes being shown. As this coil rotates from point 1 through a few degrees, it is largely sliding along the flux rather than cutting it. Jump ahead now to point 3, where a small angular change represents almost entirely effective cutting of flux. As the flux lines are shown horizontally, it can be seen that only the vertical component of motion is effective in cutting flux, and the vertical component is the sine of the angle turned through, starting from zero at position 1. Hence the generated voltage is a sine wave. At position 3 it is a maximum, say 100 v. Midway between positions 1 and 3 (at 45°), the voltage will be equal to the sine of 45° or 70.7.

A complete sine wave of 360° represents 1 cycle. If this loop is rotating 60 times per second or 3600 revolutions per minute (rpm), it will generate

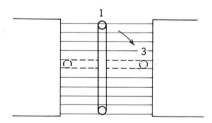

Figure 2-4. At point 1 the rotating coil is largely sliding along the flux. At point 3 it is more effectively cutting the flux.

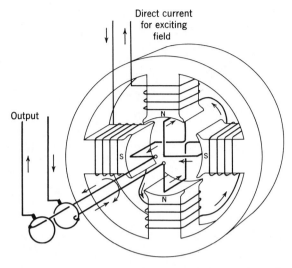

Figure 2-5. This four-pole generator needs an armature rotation of only 180° to produce a complete cycle.

a 60-cycle voltage. [A cycle per second (cps) has recently been named a hertz (Hz).]

Hence we see that the frequency of our ac power supply is dependent on the speed of the generator at the central station.

Figure 2-5 shows an elementary ac generator with four poles, alternately north and south. As the armature coils move past one north and one south pole, the voltage will pass through one cycle. Completing one revolution will generate two cycles, so to obtain a 60-cycle voltage it is necessary for the coils to turn at only half of the former speed, or 1800 rpm. In general:

$$\text{frequency, cps} = \frac{\text{pole pairs} \times \text{rpm}}{60} \qquad (2\text{-}1)$$

Note that in the two-pole generator the voltage went through 360° of the sine wave while rotating 360° in space. In the four-pole generator the voltage went through 360° of its cycle in rotating only 180°. We say in this case that 360 *electrical degrees* equal 180 *mechanical degrees*.

2-4   AC Voltage Applied to a Resistance

If an alternating voltage is connected to a resistance circuit, the rules applying to direct current still hold. As the resistance is considered con-

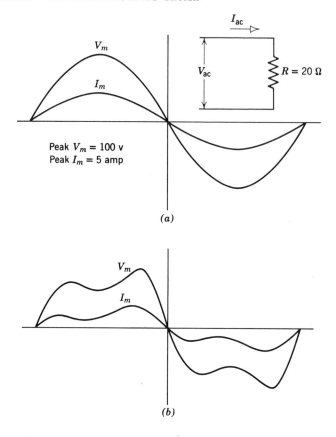

(a)

(b)

Figure 2.6. If an alternating voltage is applied to a resistance, the current that flows takes up the same shape as the voltage. At each instant i must equal V/R.

stant and the voltage varies, naturally the current will vary in the same manner as the voltage. This is illustrated in Figure 2-6. It can be seen also that, if the voltage is not a sine wave, the current will still follow the shape of the voltage. Note in part (a) that the current and voltage reach their peaks at the same time. We say that the two are *in phase*.

2-5 Effective Values

It is important that these waves be identified by some numerical value. Let us say that the maximum value of the voltage wave is 100 v. Bear in mind that it is not always at that value, but only for an instant.

If we call this 100 v and the current wave has a peak of 5 amp, the power, as the product of volts and amperes would apparently be 500 w. But a part of the time the power is zero, at other times it is somewhere between 0 and 500 (see Figure 2-7). If we identified these waves by the maximum values, what would be calculated as 500 w in a given resistance would not produce the same heat as 500 w produced by a steady 100 v dc and 5 amp dc. It is necessary that 1 w should mean the same thing regardless of the source of power. Therefore we shall identify these sine waves by a certain percentage of their peak value, so selected that

$$V \text{ revalued} \times I \text{ revalued} = \text{watts (ac or dc)}. \qquad (2\text{-}2)$$

The required value is 0.707 times the peak. Therefore this wave of a 100-v peak will be called 70.7 v. A meter across the lines would read that value. The effective current is 5×0.707 or 3.535 amp. A meter in the line would read that. Most meters are designed to read effective values.

The true average power in the circuit shown is

$$70.7 \text{ v} \times 3.535 \text{ amp} = 250 \text{ w}.$$

This is obviously the average of the 500-w peak and zero.

How was the value 0.707 obtained as a corrective multiplier? It is not the average height of a sine wave. That is 0.636 times the peak.

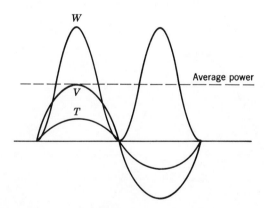

Figure 2-7. Multiplying instantaneous values of current and voltage gives the instantaneous watts. The product of two sine waves is always a sine wave. The product of two negative values is positive, and so the sine wave of watts is entirely above the zero line. The average power is the dotted line which is the center or axis of the watt wave. Note that if current and voltage are from a 60-cycle supply, the watts pulsate 120 times per second, or in general, at twice the frequency of the supply.

Power depends on I^2R or $V \times I$. Both of these represent sine squared waves. So if we square a sine wave and take the average height, and then take the square root of this average, we obtain the *effective value* of the wave. It is, as previously pointed out, 70.7% of the peak.[1]

We can see that we have three values with which we deal in considering current or voltage. These are the maximum or peak (I_m or V_m), the instantaneous values (i or v), and the effective values (I or V). Once the fundamental ideas have been developed, we shall use the simple I and V symbols and they will always refer to effective values.

2-6   Induction

We had an introduction to the self-inductance of a coil when direct current was applied and found that a countervoltage was built up only when the flux and the current were changing. But now in considering an alternating voltage and an alternating current, the flux will be constantly changing and a counterelectromotive force or voltage will be consistently present, although varying in magnitude.

Refer to Figure 2-8. At point a a line is drawn which makes an equal angle with both sides of the curve. This is the slope of the curve

[1] This is also called the root-mean-square value, so named because of the process gone through to obtain it. It is abbreviated as rms. The 70.7% factor holds only for sine waves.

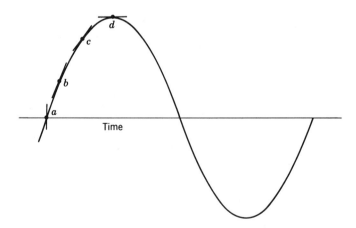

Figure 2-8. This curve may represent voltage, current, or flux. A straight line drawn tangent to the wave at any point represents the slope or rate of change at that point.

at that point. The same procedure is followed in drawing the lines at
b, c, and *d.* The slope is the rate of change. Imagine that the first
quarter of this curve is the contour of a hillside being climbed by a
man. At point *a* his rate of change in elevation is greater than at *b*
or *c,* whereas on reaching the top there is a *point* at which he walks
on level ground with no change in elevation. Considering this as a sine
wave of flux, it can be seen that the greatest *rate of change* occurs
when its value is zero, with intermediate rates up to point of maximum,
where the rate of change is zero, even though the value of flux is a
maximum.

Refer now to Figure 2-9. Here is a coil with an alternating voltage
applied. We will assume that the wire of the coil is large enough so
that the resistance is negligible. A current flows through the coil and,
if we considered only Ohm's law for resistance, the current would be
infinite. We will assume for the moment that it is a reasonable value.
The current builds up a flux which is in phase with it. Both reach their
peaks at the same time.

As the flux builds up, reduces, and repeats this in the opposite direction, at what points is the flux *changing at the maximum rate?*

The obvious answer from what we have just learned is: when it goes
through zero. By the same token the induced voltage will be zero when
the flux is a maximum value. For it is the rate of cutting, or the rate
of change in linkage, which generates or induces a voltage.

We can now draw in the position of the induced voltage, as shown
by the sine wave *e* in Figure 2-9(*b*). It is 90° out of phase with the
current and must be equal to and opposite from the applied voltage
shown as *v.* The current lags the applied volts by 90°.

In a way, the foregoing phenomenon has been described somewhat
backwards in starting with a current wave. It might be more correct
to say that, when an alternating voltage is applied to an inductance,
in order for the induced voltage to be equal and opposite to that applied,
the current takes up the necessary position and amplitude to bring this
about. Strange as it may seem, the current is a maximum when the
applied volts are zero.

Without derivation, we can show the expression for induced voltage,
thus:

$$V = (2\pi f L) \times I, \tag{2-3}$$

where $2\pi = 6.28,$
$\quad f =$ frequency in cycles per second,
$\quad L =$ inductance of the coil in henrys,
$\quad I =$ effective current.

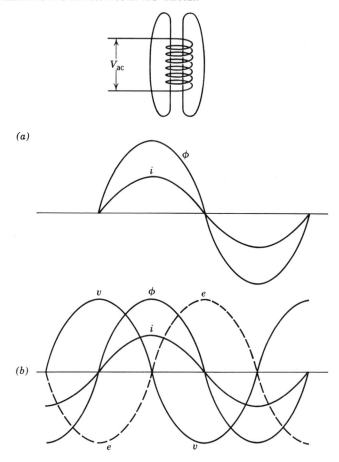

Figure 2-9. On a coil of negligible resistance the flux and current are in phase. The induced voltage, as a counter electromotive force must reach its peak when the current and flux are zero, since that is the point of maximum change.

The right-hand side of this equation shows the factors that make up the value of induced voltage; or by simple algebra

$$I = \frac{V}{2\pi f L}. \tag{2-4}$$

On the sine waves shown, one cycle represents 360°. Angles are also measured in radians (rad) as well as degrees, and 2π rad equal 360°. Then $2\pi f$ is a measure of angular velocity, which is, of course, one factor in determining the rate of cutting or the change in linkages.

2-7   Inductive Reactance

In dc circuits we know that V equals IR or, more generally, volts equal amperes times ohms. By *comparison*, then, we can measure $2\pi fL$ in ohms. It will be abbreviated X_L and is called *inductive reactance*.

Here is a very useful approach. With this concept we can make use of the simple mathematics of dc circuits:

$$V = IX_L \tag{2-5}$$

$$I = \frac{V}{X_L} \tag{2-6}$$

$$X_L = \frac{V}{I} \tag{2-7}$$

Of course, it is entirely artificial to consider that $2\pi fL$ represents ohms. Nevertheless, the system works if used carefully. The big difference between this and resistive ohms is that the inductive reactance not only fixes the value of the current but also moves it 90° out of phase with the applied voltage.

2-8   Power in an Inductive Circuit

In dealing with an alternating voltage applied to a resistance, both the voltage and current were sine waves and in phase with each other. Multiplying instantaneous values of current and voltage (i times v) resulted in another sine wave of watts of double frequency, but all above the horizonal axis. This was all positive power.

Figure 2-10 shows the result of multiplying two sine waves displaced by 90° to obtain the sine wave of power. As the instantaneous values of current and voltage are multiplied, it must be considered that $(+)$ times $(-)$ equals $(-)$, and that $(-)$ times $(-)$ equals $(+)$. The result is again a sine wave of double frequency, but equally above and below the axis. In short, the circuit takes power, kicks back an equal amount, and the net power expended in the circuit is zero. Here, again, is a peculiar phenomenon to deal with. An ammeter reads, say, 10 amp, the voltmeter across the coil reads 100 v. A wattmeter connected in the circuit would read zero. By our consideration of dc and of ac resistive circuits, the power should be 1,000 w. It is zero. It is obvious that a new factor must be introduced into power calculations.

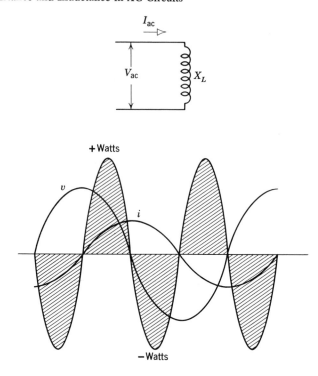

Figure 2-10. Current and voltage waves 90° out of phase result in the sine wave for watts equally above and below the axis. In such a circuit the inductance coil takes power during one-quarter cycle and kicks back the same amount during the next quarter cycle. The power wave is double frequency.

2-9   Vectors and Phasors

In dealing with ac circuits, whether they involve power distribution systems or motors, it would obviously be a great handicap to have to draw sine waves, note their relative positions, and plot out power waves, etc., to obtain useful results. We now consider a simpler approach to the problem.

Figure 2-11 shows a diagram of forces acting upon a point. These might pertain to forces on a structure or to wind velocities, etc. The resultant of these forces is obtained from completing the figure with parallel lines and drawing the diagonal. The diagonal can be scaled or calculated as to its length.

In the case shown 20- and 40-lb forces do not add up to 60 lb, because they are not operative in exactly the same direction.

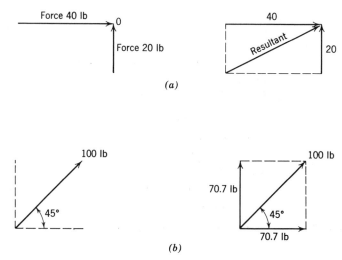

(a)

(b)

Figure 2-11. Diagrams showing how forces with both magnitude and direction can be resolved into one resultant. Similarly, a force can be broken down into components whose sum equals the original force. (*a*) Forces of 40 and 20 lb. are working on a point zero. The resultant force is the *vector sum* of the two, taking account of their direction in space. It is the square root of the sum of the squares of 44.8 lb. (*b*) A single force of 100 lb., acting at an angle of 45° with the horizontal or vertical axis, is equivalent to two forces at right angles. Each will be 70.7 lb.

That is what distinguishes these quantities from ordinary arithmetic quantities. They have direction as well as "size." Having both, they are called *vectors.* They may be out of phase *in space.*

Note in this figure that we can also take one vector and break it down into two components.

Think back now to our alternating currents and voltages. They have magnitude (we will use their effective values) and they are angularly displaced from each other *in time.* Since one cycle always represents 360°, if a current and voltage are out of phase by 90°, this is one-quarter cycle. If 60-cycle currents are considered, one-quarter cycle is $\frac{1}{240}$ of a sec.

Quantities that are displaced *in time* can be dealt with exactly like the space diagrams of forces previously mentioned, but they will be called *phasors.*

Now refer to Figure 2-12 and note the phasor diagrams for resistive and inductive reactive circuits.

It is understood that these phasors are rotating one revolution per cycle. We have stopped them at a convenient instant and given them

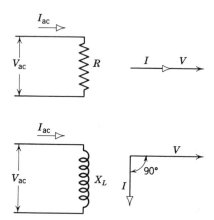

Figure 2-12. Phasor diagrams of simple resistive and inductive circuits.

a magnitude of their respective effective values. This is much easier than drawing sine waves.

2-10   Resistance and Inductive Reactance in Series

Let us consider the circuit of Figure 2-13. The inductance is "pure," having no resistance in its coil. The resistance is "pure," having no inductance. Neither of these are exactly achieved in practice.

Even though X_L is 6 Ω and R is 8 Ω, these cannot be added as being 14 Ω. They cannot be added directly because of their directional effects. R keeps the current in phase, X_L makes it lag by 90°.

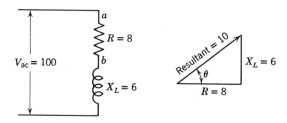

Figure 2-13. The resultant of resistance and inductive reactance is a phasor, called impedance (Z).

Note the diagram showing the two at right angles. The combined effect is measured by the diagonal.

$$\text{resultant ohms} = \sqrt{R^2 + X_L{}^2}$$
$$= \sqrt{8^2 + 6^2} \text{ or } 10. \qquad (2\text{-}8)$$

Now the total effective opposition to the flow of current is 10 Ω. It is neither resistance nor reactance, but some of both. A new word is needed for this. *Impedance* is the combined effect of resistance and reactance, taking into account the phase relationships of each. It is measured in ohms and abbreviated Z.

$$Z = \sqrt{R^2 + X^2} = 10 \text{ Ω (in this example).} \qquad (2\text{-}9)$$

$$I = \frac{V}{Z} = \frac{100}{10},$$

$$V = IZ, \qquad (2\text{-}10)$$

$$Z = \frac{V}{I}. \qquad (2\text{-}11)$$

These must be applied to the correct parts of the circuit.

To return to Figure 2-13. The current of 10 amp gives an *IR* drop across the resistance (which could be measured by a voltmeter from *a* to *b*) of 10 × 8 or 80 v. The *IX_L* drop would be 10 × 6 or 60 v. The *IZ* drop would be 10 × 10 or 100 v, exactly equal to the applied voltage. The completed phasor diagram showing these components is pictured in Figure 2-14.

Note that in figuring impedance, 6 and 8 Ω added up to 10 Ω. In considering voltage drop across the resistance and reactance, they were found to be 80 and 60 Ω. Yet they add up to 100 Ω. The idea must be grasped that, in dealing with sine waves, vectors, or phasors, 2 plus 2 is not always 4. The sum can be anything from 0 to 4, because sine waves, vectors, or phasors may have direction (or displacement).

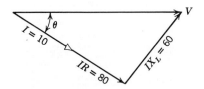

Figure 2-14. The complete phasor diagram for the circuit of Fig. 2-13 shows that 80 v are dropped across the resistance and 60 across the inductive reactance. As phasors they equal the 100 v applied.

2-11   Power in a Circuit

In the resistance circuit we found that power (based on effective values of current and voltage) was exactly the same as in dc calculations. In the example

$$\text{power in } R = I^2R$$
$$= 10^2 \times 8 = 800 \text{ w,}$$

power lost in a pure inductance is zero.

The only power in this circuit is that of the resistance, or 800 w. Yet if we look at the input of 10 amp and 100 v, the apparent power is 10×100 or 1000. We shall call it exactly that, *apparent power*. As it is the product of volts and amperes, it will be measured in *volt-amperes* (va) or in kilovolt-amperes (kva). The latter is 1,000 va.

Here is an apparent power of 1000 va and an actual power of 800 w. It is obvious that we must multiply apparent power by some factor, always one or less to obtain real power. This is called the *power factor* (pf). Thus

$$\text{power} = V \times I \times \text{pf,} \tag{2-12}$$

or

$$\text{pf} = \frac{\text{watts}}{\text{va}}. \tag{2-13}$$

In a pure resistance circuit, pf = 1. In a pure inductive circuit, pf = 0.

Figure 2-14 shows the phasor diagram for the foregoing example, indicating that, of the 100 total volts, one component is completely in phase with the current, and the other component is 90° out of phase. These components represent the cosine and sine of the angle by which the current lags the voltage. Therefore the power factor is also the cosine of the angle between current and voltage. Also

$$\text{pf} = \cos \theta = \frac{R}{Z}. \tag{2-14}$$

2-12   Resistive and Inductive Reactance in Parallel

In the examples considered the circuit elements have been in series, with a current common to both but with different voltage drops. In the parallel circuit the same voltage is applied to R and X_L, but each take an individual current.

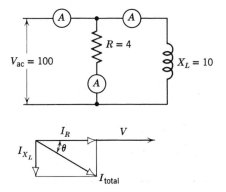

Figure 2-15. Resistive and inductive reactances in parallel.

The simplest approach is first to determine the current in each branch (see Figure 2-15).

$$I_R = \frac{V}{R} = \frac{100}{4} \text{ or 25 amp, in phase.}$$

$$I_L = \frac{V}{X_L} = \frac{100}{10} \text{ or 10 amp, lagging.}$$

$$I_{\text{total}} = \sqrt{I_R{}^2 + I_L{}^2}$$
$$= \sqrt{625 + 100}, \text{ or 26.9 amp.}$$

The combined impedance of these two in parallel must be

$$Z = \frac{V}{I} = \frac{100}{26.9} \text{ or 3.72 } \Omega,$$

$$\text{power} = I^2R = 25^2 \times 4 \text{ or 2500 w,}$$

$$\text{apparent power} = \text{va} = 100 \times 26.9 \text{ or 2690 va,}$$

$$\text{pf} = \frac{2500}{2690}, \text{ or 0.935.}$$

In series circuits, the power factor also equaled R/Z. This would be 4/3.72 or 1.075. This being greater than unity, is an impossibility. The ratio R/Z is not the power factor in parallel circuits. It can be calculated as I_R/I_T.

Chapter 3

The Capacitor,
Further Circuit Calculations

3-1 The Capacitor

Figure 3-1 shows two metal plates separated by a film of insulation. If these plates are connected to a source of direct current, at the instant the switch is closed the ammeter in the line will show that a current flows. If the voltage is reduced, the current flows in the opposite direction.

Here is a very peculiar thing. One part of the circuit is insulated from the other part, and yet, momentarily at least, a current flows seemingly as though they were connected. This is a phenomenon of the dielectric circuit or dielectric field. (We have already discussed the electric and magnetic circuits.) Dielectrics, loosely, are insulators. We shall ignore this subject except for this one application. The voltage on this metal plate piles up electric charges in it; to do so, electrons move into the plate. The charges also cause an equal movement of electron flow out of the other plate. As the movement of electrons in a conductor is a current, we have a current flow measured by the ammeter. No electrons flow from one plate to the other. These charges continue to build up until the difference in voltage between the two plates is exactly equal to the applied voltage. Remove the source of voltage and the capacitor retains its voltage, for a minute or for several hours. It is "charged."

This device used to be called a *condenser*, but because a completely different device used in refrigeration and steam power plant work is also called a condenser, to avoid confusion, the electric condenser is now called a *capacitor*.

It is important to note that current flows only when the voltage is changing. The amount of current depends upon the area of the plates,

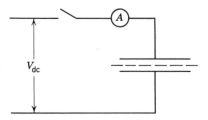

Figure 3-1. An elementary capacitor consisting of metal plates and an insulating separator is connected to a source of direct current.

the distance between them, and the type of insulating material separating the plates.

The capacity is measured in *farads* (f). A capacitor of 1 f will take a current of 1 amp if the voltage changes at the rate of 1 v/sec. It is too large a unit to be practicable, so the term used commonly is microfarad (μf) which is 1/1,000,000 of 1 f.[1]

The modern capacitor usually consists of two long strips of thin aluminum foil separated by an insulating film of special paper or plastic. These strips are made into a tight roll and sealed with oil into a metal or plastic can out of which extend two terminals. The size depends somewhat on the voltage at which the capacitor is to be used. See Figure 3-2 for a hydraulic analogy to the capacitor.

3-2   Capacitors and an Alternating Voltage

In considering charging a capacitor with direct current, it was found that there is a movement of electrons in or out of the plates (current

---

[1] The forerunner of the capacitor was the Leyden jar, invented at the University of Leyden in 1746. This was usually in the form of a glass tumbler with a metal sheet on the inside and another on the outside. By rubbing silk on a glass rod and touching the glass to the inner cup, a charge was built up. Repeating this several times increased the charge. Touching the inner and outer cups could result in a severe shock.

Benjamin Franklin played with this and saw the possibility of replacing the glass tumblers or bottles with flat sheets of glass with foil between. He therefore built the first modern capacitor.

Society was intrigued with this novel device. Parties were held in Europe, Britain, and America in which the guests would join hands while the "end men" touched the respective terminals, causing the shock to pass through the entire group. It was recommended as titillating the nervous system and speeding the flow of body juices.

Experiments of this sort led Franklin to suspect that lightning was an electric discharge. A few years later he flew his kite and developed the lightning arrester.

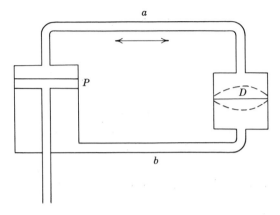

Figure 3-2. Hydraulic analogy to the capacitor. The piston P moves up and down in the cylinder, forcing water through the pipes with each movement. An elastic diaphragm D bulges in response to the piston movement. Water flows in the pipes as a result, but no water ever flows from a to b or vice versa.

flow) only when the voltage is changing. Apply an alternating sine wave voltage to this device, and current of a sine wave shape will flow continuously. But the rate of flow will depend on the rate of change of voltage. We found in examining sine waves that the *rate* of change is always greatest when the wave goes through zero, and is zero when the wave is at its peak.

Therefore a capacitor takes its greatest charging current when the voltage is zero. The current and voltage are 90° out of phase, and the current must lead the voltage (see Figure 3-3).

Mathematically the current is

$$I = V(2\pi fC), \tag{3-1}$$

where $2\pi f$ is the expression for angular velocity as used before, hence is a measure of the rate of change, and C is the "size" of the capacitor in farads.

It is readily understood that an increase in voltage, frequency, or in the size of the capacitor would all affect the current flow.

3-3 Calculations

In dealing with inductance we were able to simplify our calculations by taking the terms $2\pi fL$ and considering this expression as ohms, so

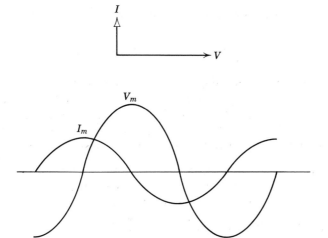

Figure 3-3. Relationship of voltage and current in a capacitive circuit.

that circuit solution was similar to Ohm's law in direct current. With capacitors we can do the same thing. However, we cannot call $2\pi fC$ "ohms" because amperes are equal to volts *divided* by ohms, not multiplied by them. The term $2\pi fC$ can be considered as ohms "upside down."[2]

Therefore we shall write

$$\text{ohms} = \frac{1}{2\pi fC},\tag{3-2}$$

where C is in farads, which is too large a unit.

$$\text{ohms} = \frac{10^6}{2\pi fC} = X_C,\tag{3-3}$$

where C is in microfarads and X_C is called the *capacitive reactance*.

A capacitor of 400 μf is used on a 60-cycle circuit. What is the capacitance reactance?

$$\frac{10^6}{6.28 \times 60 \times 400} = 6.62 \ \Omega.$$

[2] The term for this is *mho*($\mho$), which is ohm spelled backwards. Aside from this bare mention of the fact we shall avoid this complication. Students who hear this term used the first time have never quite been able to decide whether this displays a sense of humor, or a lack of it, on the part of electrical engineers.

With 115 v applied, how much current will flow?

$$I = \frac{V}{X_C} = \frac{115}{6.62} \text{ or } 17.35 \text{ amp (leading).}$$ (3-4)

As this current is out of phase with the voltage by 90°, the sine wave of watts is equally above and below the zero line. True power is zero. During one-quarter cycle the capacitor takes power from the line, in the next quarter-cycle it kicks back an equal amount.

We have now considered the third and last circuit element.

Resistance causes the current to be in phase with the voltage.
Inductive reactance causes the current to lag the voltage.
Capactive reactance cause the current to lead the voltage.

Circuits can be made up of any combination of these elements.

3-4   R and X_C in Series

A resistance of 6 Ω is in series with a capacitive reactance of 15 Ω, as shown in Figure 3-4. Determine the current, the power factor, and the power in the circuit with 115 v applied.

The impedance Z is again the combination of R and X_C, taking into account their effect in the circuit. Adding them 90° out of phase gives

$$Z = \sqrt{R^2 + X_C{}^2}$$

$$= 36 + 225 \text{ or } 16.19 \ \Omega.$$ (3-5)

$$I = \frac{115}{16.19} \text{ or } 7.14 \text{ amp.}$$

$$\text{pf} = \frac{R}{Z} = \frac{6}{16.19} \text{ or } 0.371 \text{ (leading).}$$ (3-6)

$$\text{power} = V \times I \times \text{pf}$$

$$= 115 \times 7.14 \times 0.371$$

$$= 305 \text{ w.}$$

The voltage drop across the capacitor is

$$IX_C = 7.14 \times 15 \text{ or } 107 \text{ v.}$$

The voltage drop across the resistance is

$$IR = 7.14 \times 6 \text{ or } 42.9 \text{ v.}$$

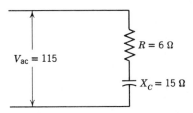

Figure 3-4. A resistance and capacitance in series. The impedance is a phasor of 16.19 Ω.

3-5 *R*, X_C, and X_L in Series

Refer to the circuit of Figure 3-5. We wish to find the current in the circuit, the voltage across each element, the power, and power factor.

It is necessary to determine the impedance by considering the effect of each element on phase position. Since X_L makes the current lag and X_C makes it lead, the two are directly opposite in their effect and can be subtracted.

$$X_{net} = X_C - X_L$$

$$= 12 - 8 \text{ or } 4 \text{ Ω (leading)}. \tag{3-7}$$

$$Z = \sqrt{R^2 + (X_{net})^2}$$

$$= \sqrt{3^2 + 4^2} \text{ or } 5 \text{ Ω}. \tag{3-8}$$

$$I = \frac{V}{Z} \text{ or } \frac{100}{5} \text{ or } 20 \text{ amp.}$$

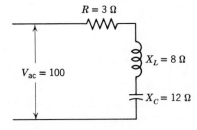

Figure 3-5. Resistance, capacitance, and inductance in series. X_L and X_C are directly opposite in their effect and must be subtracted to find the net reactance, hence the impedance.

Volts across the elements:

$$IR = 20 \times 3 \text{ or } 60 \text{ v,}$$

$$IX_L = 20 \times 8 \text{ or } 160 \text{ v,}$$

$$IX_C = 20 \times 12 \text{ or } 240 \text{ v.}$$

$$\text{true power} = I^2R = 400 \times 3 \text{ or } 1200 \text{ w.}$$

$$\text{apparent power} = VI \text{ or } 100 \times 20 = 2{,}000 \text{ va.}$$

$$\text{pf} = \frac{\text{watts}}{\text{va}} = \frac{1{,}200}{2{,}000} \text{ or } 0.60 \text{ (leading)}$$

$$= \frac{R}{Z} = \frac{3}{5} \text{ or } 0.60 \text{ (as above).}$$

Note that with only 100 v applied, voltages across parts of the circuit are 60, 160, and 240. Two of these voltages are actually higher than the supply. This can occur in certain types of motors.

3-6   Series Resonance

In the foregoing circuit, we have seen how the net effect of X_C and X_L always results in a reduced quantity because they neutralize each other. Suppose now that the inductive reactance of X_L were increased. Ordinarily we would consider this as *reducing* the current that could flow, but if X_L is 12 Ω,

$$X_C - X_L = 0.$$

The total impedance reduces to R of 3 Ω.

$$I = \frac{V}{R} = \frac{100}{3} \text{ or } 33.3 \text{ amp,}$$

$$IR = 100 \text{ v,}$$
$$IX_C = 33.3 \times 12 \text{ or } 400,$$
$$IX_L = 33.3 \times 12 \text{ or } 400.$$

The power is I^2R or $33.3^2 \times 3 = 3330$ w. The power factor is unity.

This peculiar condition is known as series *resonance*. Under the condition of resonance the power in a circuit is the maximum for any given value of resistance, the current is a maximum, and the voltage across the elements is a maximum.

Internally, during the quarter-cycle in which the capacitor is taking power, an equal amount is being kicked back by the inductance, and vice versa. Thus, although there is an exchange of power at a high level within the two, it is not reflected into the supply circuit. An almost similar condition can occur in some electric motors.

3-7 R, X$_L$, and X$_C$ in Parallel

One more example is presented of a combination of circuit elements. This involves the parallel circuit shown in Figure 3-6. Obviously, the current is different in each branch and the voltage on each is the same, neglecting any resistance in the leads. We solve for the currents:

$$I_R = \frac{V}{R} = \frac{100}{50} \text{ or 2 amp in phase.}$$

$$I_C = \frac{V}{X_C} = \frac{100}{20} \text{ or 5 amp leading.}$$

$$I_L = \frac{V}{X_L} = \frac{100}{20} \text{ or 5 amp lagging.}$$

Note the phasor diagram indicating that the current in the inductance and the current in the capacitor are exactly equal and opposite. The

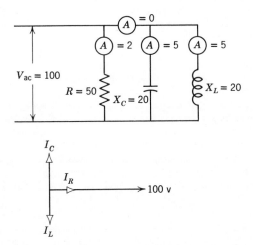

Figure 3-6. Resistance, capacitance, and inductance in parallel. The fact that X$_C$ = X$_L$ puts this circuit into parallel resonance by which larger currents can flow in the branches than are fed into the circuit.

current in the line supplying these two elements is zero, even though 5 amp are circulating around this part of the circuit. The circuit as a whole takes only 2 amp or 200 w at unity power factor. This circuit is in *parallel resonance*.

In all of these examples it has been assumed that a 60-cycle voltage is applied. In the resonant circuits shown with elements in series or parallel, practically *any* combination of X_L and X_C can be put into resonance by varying the frequency. This is possible because X_L goes up with frequency increase, whereas X_C goes down under the same change. At some point they become equal, and a maximum response is obtained in the circuit. Everytime you tune your radio you are changing X_L or X_C so that it is in resonance for the station you desire.[3]

---

[3] In the early days of alternating current the solution of circuits required differential equations capable of dealing with quantities that are constantly changing. Later on, Steinmetz developed the mathematics for ac circuits as we know them today by greatly simplifying calculations. It is because of him that we can write the comparatively simple equation

$$V = IR + IX_L - IX_C.$$

These, of course, are phasors, not arithmetic sums. It is alleged that Steinmetz developed most of his lifework in the mathematics of circuits and machinery through a simple statement in one of Oliver Heaviside's technical papers. Oliver Heaviside (1856–1925) was an English recluse living in Torquay whose one hot meal per day was delivered by the local policeman. He wrote obtuse papers on electrical physics for the *London Electrician* and reputedly became embittered because they did not immediately receive the attention they deserved. It was he who predicted that radio (then wireless) was possible because the waves bounced off an ionized layer in outer space. He gained some recognition by having this Heaviside layer named for him. He also suggested the name *inductance*, but it was Steinmetz who coined the word *reactance*. Although the solution of ac circuits was still in the differential-equation stage, it was Michael "From Immigrant to Inventor" Pupin, a professor at Columbia University, who solved the equation for resonance. In working with wave propagation through space and over wires, he recognized that two long conductors as represented by telephone or telegraph lines had a capacitor effect causing distortion. To obtain a suitable resonant balance, he calculated the size and placement of inductance along the lines. These are known as *loading coils*, and Pupin, who was not especially a modest man, tried to have this process called "Pupinizing." It was never accepted. However, he sold his patent rights to the American Telephone and Telegraph Company and said that he "was treated most generously." It was reputed to involve $1,000,000. He made long-distance telephony possible.

Resonance is not an isolated electrical phenomenon. Every thing in nature which has elasticity (corresponding to capacitance) and weight or inertia (corresponding to inductance) has a natural frequency, calculated by equations which have a similarity to those for electric resonance. Perhaps the reader has stood in church with his hands on the pew ahead of him and noted that at certain notes from

3-8   General Fundamentals now Completed

It may occur to the reader at this point that he is still far away from an understanding of motors and their application. The preceding material has not been merely mental gymnastics, introduced for its own sake. All ac motors are made up of combinations of resistance and inductive reactance, and some include capacitive reactance as well. A much better picture as to what occurs in motors comes from an understanding of the material presented in the foregoing chapters.

---

the organ the pew vibrated. The frequency of that note agreed with the resonant frequency of the pew, and it responded. Although Pupin acknowledged that the ideas of Heaviside aided him in his developments, he also told that the shepherds in the hills around the Serbian village in which he was raised would drive their daggers into the ground and signal each other by one shepherd "twanging" the handle and the other, at some distance, receiving the message by putting his ear to the handle of his dagger. This was resonance, and he sought the electrical equivalent of it.

Here are at least a few of the later scientists who made possible the material presented in the past few chapters. Some made money, others did not, and their fame was not long-lasting. Few of the younger engineers of this day even know their names.

Chapter 4

One-, Two-, and Three-Phase Generation, Transformers and Meters

4-1 Single-Phase Generators and Windings

In Chapter 2 we saw an elementary ac generator (alternator) consisting of a loop of wire turning at uniform speed in a magnetic field of uniform flux density. This resulted in a sine wave of voltage being generated in the loop. As such it is impractical. The large air gap between the poles would make it difficult to build up a reasonable flux density, and the single loop of wire would produce only a low voltage.

A better way to build an alternator is shown in Figure 4-1. Here the dc field actually rotates with direct current fed to coils on each pole by brushes and slip rings. Any even number of poles can be used, depending upon the intended speed of the alternator and the frequency desired. As shown, operation at 1,200 rpm would produce 60 cycles.

Note the stationary armature winding.[1] The concentrated coils are connected in such a manner that the voltages in each add up to the final output voltage. A variable resistance in the dc supply line can be changed to raise or lower the excitation current, thereby varying the flux in the gap and the alternating-voltage output.

As shown, this alternator is still not constructed in its most effective form. Note that the winding for each pole is concentrated in one slot. This does not make the most use of the periphery of the armature. Commercial machines usually have many slots with the windings divided up into multiple coils. Figure 4-2 shows three types of windings com-

---

[1] Some confusion exists here in nomenclature. The dictionary definition of armature implies a moving part. But when alternators were "turned inside out," so to speak, the armature became the stationary member.

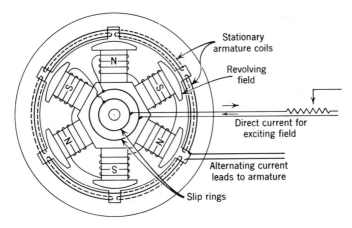

Figure 4-1. A more practical method of constructing an alternator in which the dc field rotates. The air-gap flux cuts the stationary ac winding.

monly used. It is obvious that the voltages in coils 1, 2, and 3, for instance, cannot all be a maximum at the same time. The sine waves of each set of coils under one pole pair are out of phase with each other by an angle determined by the electrical degrees between the slots.

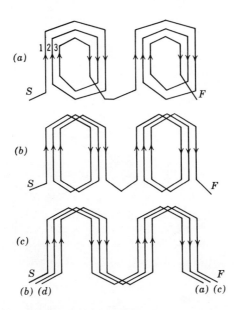

Figure 4-2. Elementary single-layer, single-phase windings. (*a*) Spiral. (*b*) Lap. (*c*) Wave.

The output voltage of distributed windings is, therefore, slightly less than the arithmetic sum of the volts of each coil.

4-2   Applications

Large single-phase alternators are not built and sold in great quantities. Perhaps the largest sales are in the 1- to 5-kva range. These alternators are used as auxiliary sources of power, driven by tractors, making electric power available for portable tools or for household supply in case of main power failure.

Gasoline-engine-driven single-phase alternators built as a unit are quite popular as portable or standby power. A farmer with, say, 500 lb of meat in his "deep freeze," fearing a "blackout," is a likely customer.

4-3   Two-Phase Generation

Let us return now to the elementary alternator with a loop of wire rotating in a magnetic field. In Figure 4-3 a second loop has been added with terminals connected to additional slip rings. These loops are 90° apart (in a two-pole machine). When the voltage of coil 1 is a maximum, that of coil 2 is zero, and vice versa. The waves are shown in Figure 4-4. This, by definition, is a two-phase alternator. The alternator

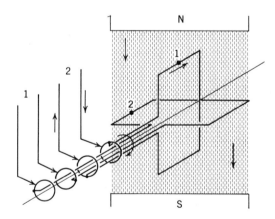

Figure 4-3. A second loop of wire has been added, displaced 90° from the original loop. The voltage in each circuit reach their maximum values 90° or ¼-cycle apart. This is a two-phase system.

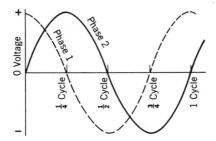

Figure 4-4. The alternator of Figure 4-3 generates two voltage waves displaced as shown.

Figure 4-1 could be adapted for two-phase operation by placing a second winding in slots midway between those shown.

When the power industry was young, some supply systems were built based on two-phase generation. As we shall see in dealing with motors, the two phases simplify motor construction over the single phase but, as shown in Figure 4-5, the system results in an odd, nonstandard voltage

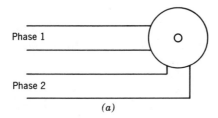

(a)

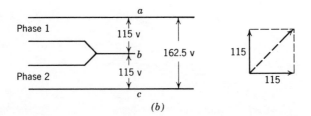

(b)

Figure 4-5. A four-wire and a three-wire two-phase distribution system. (a) Two phases supplying a two-phase motor. (b) A two-phase system combines into a three-wire distribution system. Lines a–b and b–c can be loaded, but lines a–c show 162.5 v between them. This is not generally useful. It results from the sum of 115 and 115 at right angles to each other.

between outside lines, which is not generally useful. The Philadelphia area is the last district in which two-phase power distribution is still used.

4-4   Three-Phase Generation

Figure 4-6 shows an alternator in which three coils are rotated on a common shaft. They are located 60° apart or, more properly, 120° apart, depending on which terminal of the second coil is considered the *start* or the *finish*.

Six slip rings are shown. However, it is possible to connect one terminal of each of the three sets of coils to a common point (called the *neutral*) and bring out only the three remaining terminals to three slip rings. This is called a Y or star connection.

Note the four-pole construction of Figure 4-7. There are three slots per pole. The Y connection described in the foregoing is shown in Figure 4-7*b* with a wave-type winding of one coil side per slot.

The alternate connection is known as the Δ (delta) connection, illustrated in Figure 4-7*c*. This would appear to short-circuit all of the windings unless we consider the phase displacement between them.

Whether a three-phase alternator is designed to be connected internally as a Y or delta machine is of importance to the designer, but is rarely of any importance to the power user.

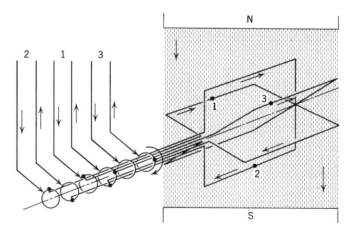

Figure 4-6. An elementary three-phase alternator with loops displaced 120°. Usually internal connections are made in the windings so that only three slip rings are needed.

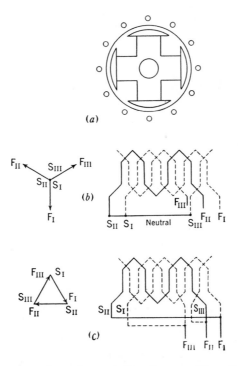

Figure 4-7. Fundamental three-phase winding diagram. Lines F_I, F_{II}, and F_{III} are brought to the external circuit V by slip rings and brushes.

4-5   Current and Voltage Relations in Three-Phase Circuits

Consider a three-phase alternator in which the three windings are Y-connected. The terminals A, B, and C are connected to three equal resistances, as shown in Figure 4-8.

Each of the three windings generates 100 v between the neutral O and the respective terminals A, B, and C. Each leg of the resistive load, OA', OB', and OC', will have 100 v impressed.

Assume the resistance of each leg is 10 Ω. Then 10 amp flow in each leg, each reaching their maximum values 120° apart as dictated by the 120° phase positions of the alternator windings.

If we could measure the power in each leg, O to A', O to B', and O to C', we would find each taking $100 \times 10 \times 1$ w. The power factor of these resistive loads is unity and the total power put out by the alternator is 3,000 w.

Suppose it were impossible or impractical to measure power in each

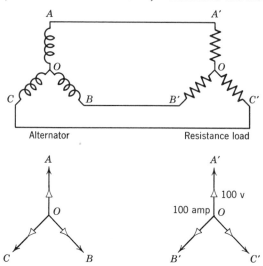

Figure 4-8. Voltage and current relationships in a three-phase Y-connected alternator and a balanced three-phase Y-connected resistive load.

of the legs. That is, connection points O are inaccessible and meters must be placed in the lines AA' or between lines A and B, etc. Now we must consider the voltages between the lines and in the individual legs.

Refer to Figure 4-9. We know that each leg voltage peaks 120° out of phase with the other windings. We can complete the figure for any two legs showing that the diagonal or sum is made up of the side of two equal triangles of 60°. The sine of 60° (ratio of vertical to the hypotenuse) is 0.866. Therefore, the phasor sum of OA and OB is 2×0.866 or 1.73 times the voltage of one leg; 1.73 is $\sqrt{3}$.

A voltmeter between any two lines would reach **173** v. An ammeter in each line would read **10** amp. It would appear, then, that the total power in the circuit is

$$3 \times 173 \times 10 \times 1 \text{ or } 5200 \text{ w.}$$

But **173** v is not the power impressed on each load. We know it to be $173/\sqrt{3}$. The true power therefore is

$$3 \times 173/\sqrt{3} \times 10 \times 1 = \sqrt{3} \times 173 \times 10 \times 1 \text{ or } 3,000 \text{ w;}$$

or, in general, power in a three-phase circuit is

$$\text{power} = \sqrt{3} \, V_\text{line} \times I_\text{line} \times \text{pf} \tag{4-1}$$

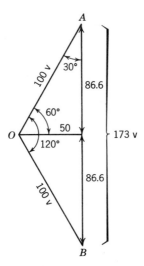

Figure 4-9. In Y-connections the voltage between lines is 1.73 times the voltage to neutral. In spite of unity-power factor resistive loads in the legs, the voltage between supply lines is 30° out of phase with the currents.

4-6   Relationships in Delta Connections

The same Y-connected alternator used in the preceding discussion is now connected to three 10-Ω resistors connected in delta (see Figure 4-10). Now 173 v are applied to each resistance and the current is increased.

$$I_{\text{in each leg}} = \tfrac{173}{10} \text{ or } 17.3 \text{ amp.}$$

The power in the three legs is

$$3 \times 173 \times 17.3 \times 1 \text{ or } 9000 \text{ w.}$$

Assume that it is not possible to connect meters in the individual legs of the circuit but only the connecting lines are available. We know that load and line voltages are the same, namely, 173 v, but the line currents are $\sqrt{3}$ times the current in the legs as indicated in Figure 4-10. Therefore, an ammeter in each line would read $17.3 \times \sqrt{3}$ or 30 amp.

Using line values, we would apparently have

$$3 \times 173 \times 30 \times 1 \text{ or } 15,600 \text{ w.}$$

Since I in the Δ leg is

$$I_{\text{leg}} = \frac{I_{\text{line}}}{\sqrt{3}} \tag{4-2}$$

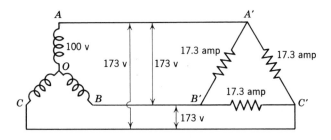

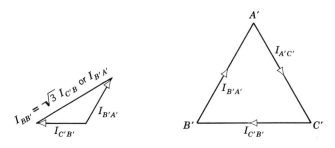

Figure 4-10. The Y-connected alternator is now connected to a balanced delta-connected resistive load, with 10 Ω in each leg. Note that although the line and load voltages are the same, the current in the supply lines is $\sqrt{3}$ times the leg current, following the same analysis used in determining the voltages of Figure 4-9. In spite of the resistive, unity-power factor load in the legs, currents in the supply are 30° out of phase with line voltages.

the real power in the load is

$$3 \times 173 \times \frac{30}{\sqrt{3}} \times 1 \text{ or } 9000 \text{ w};$$

or, in general, power in this three-phase circuit is

$$\sqrt{3}\ V_{\text{line}} \times I_{\text{line}} \times \text{pf} \qquad \text{(same as Eq. 4-1).}$$

Here is a very useful fact. In measuring power taken by a three-phase circuit, we do not need to know the connections back in the supply alternator, nor do we need to know whether the load (a motor, for instance) is connected internally Y or delta. In any case the same line measurements can give us the true power.

The novice is frequently confused by the necessity of the factor $\sqrt{3}$ in three-phase calculations. In the foregoing example, an ammeter in each line would read 30 amp, the loads were purely resistance, and certainly the power must be based on three times these **30** amp and the voltage. The catch is that at one instant **30** amp are going out through one line and return in the other two. An instant later **30** amp are flowing to the load in the second lead and return through the other two. The meters read the effective value of current in each line but deceive the observer by reading both the current going to the load as well as the return current. Hence a correction is needed, and we have seen from both the Y and delta cases that this factor is 1.73 or $\sqrt{3}$.

4-7   Transformers

The distribution of power in one-, two-, or three-phase systems depends on the transformer; also, many electric motors operate on what might be called the *transformer principle*. In Chapter 1 we saw a few statements that indicated how a transformer operates. It might be well at this point to enlarge on these principles.

Figure 4-11 shows a magnetic core built of laminated steel on which two windings are placed. The first, called the primary, is connected to the source of ac power. The second, called the secondary, supplies the load. For convenience these windings are shown on two separate legs of the core. In practice to make sure that all of the flux of one coil links with the other primary and secondary windings are usually wound as closely as possible on both legs.

The windings are of low resistance, so if the primary were connected to, say, a 440-v line, it would apparently take a very large current.

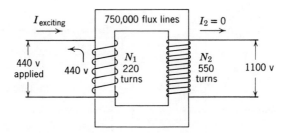

Figure 4-11. Voltage and current in a step-up transformer with no load on the secondary winding N_2. At no load the primary takes the current from the line as required to build up a flux which induces a counterelectromotive force equal and opposite to the applied voltage. We know the basic equation is $V_{\text{self-induced}} = 4.44\, fN\phi_{\text{max}}10^{-8}$. The ϕ_{max} will be $(440 \times 10^8)/(4.44 \times 60 \times 220)$ or 750,000 lines. The exciting current required to do this depends on the length of the cross section and the quality of the laminated steel core.

But as we saw in Chapter 1, the alternating flux built up in the core cuts both primary and secondary coils, giving the former a self-inductance, leading to a counter voltage that reacts against that applied.

Neglecting the slight IR drop in the primary, with the secondary lines open, the transformer takes just enough current from the line to build up a flux resulting in a counter electromotive force equal and opposite to the applied voltage. If the steel in the core were of poor magnetic quality or an air gap existed in the core, this magnetizing current would be high.

Suppose there were 220 turns of wire in the primary coil. Then 440 v are built up as the counter electromotive force to oppose the applied voltage. This is enough flux to induce 2 v per turn. Assuming that all of this flux links the secondary coils, 2 v per turn will be induced in each. If there are 55 turns on the secondary, 110 v would be the output. If there were 550 turns on the secondary, the output would be 1100 v.

Thus a transformer steps up or steps down voltage, depending entirely on its turn ratio.

When a load is applied to the secondary, say 10 amp, this current flows through the secondary coils of, say, 550 turns. This coil now has a demagnetizing force of 10×550 or 5,500 amp-turns, working against the magnetizing force of the primary, which must remain large enough to build flux to induce the proper counter electromotive force. As a result, the primary current increases until $220 \times I_1 = 5500$. Therefore the primary *load* current must be 25 amp. This is enough to counterbalance the demagnetizing effect of the secondary load current, but,

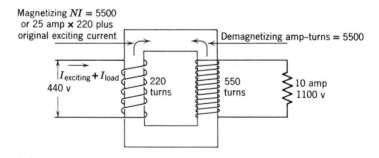

Figure. 4-12. This magnetic core shows how the presence of a load on the secondary of the transformer in Figure 4-11 produces demagnetizing effect which must be overcome by additional current drawn from the supply line. The current is added in suitable phase position to the original current required to build up the original 750,000 flux lines in the magnetic cove. The example is slightly in error because the small IR drops in voltage are neglected in both windings.

in addition, the primary also requires the original component of current needed for magnetizing (see Figure 4-12).

Then

$$\frac{V_1}{V_2} = \frac{N_1}{N_2} = \frac{I_2}{I_1}.$$ (4-3)

This expresses the ideal case; small voltage drops in the primary and secondary upset the voltage ratio slightly, and the presence of the primary magnetizing current also modifies the current ratio.

4-8 Special Transformers

There are two types of special transformers or transformer connections with which the motor user is likely to come in contact. These are (1) the autotransformer and (2) the Scott or T connection. They will be described briefly in the following:

1. The autotransformer depends for its action on a single continuous winding used for both input and output voltages. It can either step up or step down the voltage (see Figure 4-13). These windings are sometimes used with certain types of motors as a means of speed control.

2. The T connection is a method of transforming two-phase motors to three-phase motors, or vice versa. Either autotransformer or the two winding types can be used for this purpose. Figure 4-14 illustrates the later type.[2]

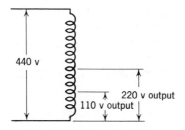

440 v

220 v output

110 v output

Figure 4-13. Schematic representation of an autotransformer with the magnetic core not shown. With 440 volts applied to the entire winding, taps can be provided to obtain any output voltage desired. Note that the supply and output voltages are not isolated in separate windings; a breakdown of the insulation could subject the output to the primary voltage. If 220 v were supplied to the taps as shown, this would become a step-up autotransformer with a 440-v output.

[2] Developed by Professor Charles Scott about 1894. In those days two and three phases were known as "diphase" and "triphase" respectively.

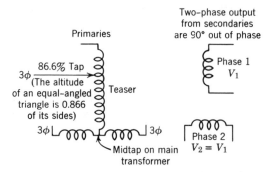

Figure 4-14. The *T*-connected transformers are shown schematically for connection to a three-phase supply. One transformer, called the "teaser," must have an 86.6% tap in its primary winding. The main must have a midtap for connection as shown.

4-9   Meters

Casual mention has been made here and there throughout this book of measuring the current in a line, the power input, etc. From a variety of meter types let us examine briefly one type that always reads the effective value of a current wave, a voltage wave, or the true power in a circuit considering the power factor. These instruments are called *dynamometers.*

1. Figure 4-15 illustrates the schematic arrangement of an ammeter with two stationary coils *S* and *S'* connected for like polarity. A moving coil *M* is suspended between them on a spindle set in jewel bearings. Terminals of the moving coil connect through springs in series with the stationary coil. Connections are such that the coil sides of the moving coil are repelled by the magnetic field of the stationary coils. As there is no iron in the circuit to saturate, the fields are directly proportional

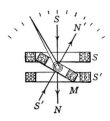

Figure 4-15. The principle of the electrodynamometer type of meter depends on the opposing magnetic forces of two coils reacting against springs.

to the current, which, being the same in both sets of coils, produces a torque proportional to i^2, working against the springs. As the needle cannot respond to the rapid fluctuations of the i^2 curve (double frequency), it takes up an average position, thereby indicating the effective current on the scales.

The use of the springs as conductors supplying the moving coil, limits the current that should be measured by such a meter. Naturally, it is connected in series with the line to measure rate of flow in the conductor. We shall see later how an instrument transformer increases its range of measurement.

2. *Electrodynamometer-type voltmeters.* If a resistance of several thousand ohms were connected in series with this instrument, it could be connected safely across a high-voltage line. Because the series resistance is constant, the current that flows through the coils is directly proportional to the voltage applied. Hence suitable calibration of the scales enables the meter to read effective voltage.

3. *Wattmeters.* Nearly all wattmeters are of the dynamometer type. Because they must measure both current and voltage, it is customary to connect the stationary coils in series with the line so that their fields are proportional to the current. The moving coil is connected in series with a resistance of thousands of ohms across the line. The reaction of the field of this coil is again proportional to the current, determined by the voltage and fixed internal resistance. If the current in the stationary coils is out of phase with the current in the moving coil (the latter being fixed by the voltage), the two magnetic fields do not react with the same torque. The pointer is, therefore, deflected less, and it can

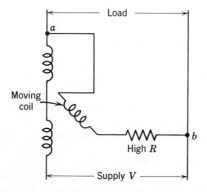

Figure 4-16. The dynamometer-type wattmeter, which reads true power in the load. However, with the voltmeter leads connected at *a* and *b*, as shown, the meter also reads the loss in the voltmeter coil and resistance. Often this is negligibly small.

be proved mathematically that this reduction is equal to the cosine of the angle between I and V. Therefore a wattmeter reads true power or $VI \times \text{pf}$.

4-10   Power Measurements—Small-Motor Testing

When a motor supplier submits a carefully tested small motor as a sample to a prospective customer, this sample is usually tested also in the customer's laboratory as a check. With exactly the same load applied, the two tests sometimes disagree. The reason is often the result of different methods of connecting the meters (see Figure 4-17 for two alternate methods). Note that neither method gives exact input. However, if all parties can agree as to the method to be used, reasonable cross checking on motor tests can be achieved.

The modern approach usually employs the connections of Figure 4-16, with the resistances of each meter recorded with the laboratory data. Then when these data are processed through a computer for obtaining complete performance, all meter "drops" or losses are eliminated by a suitable computer program.

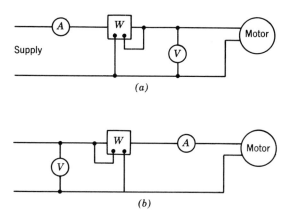

Figure 4-17. Two of several methods of connecting meters in the line for measuring motor input. In small motors, especially, these meter losses are important. (a) If the ammeter is rated as, say, ¼ to ½ amp, its resistance might be 10 to 5 Ω. This plus the resistance of the current coil of the wattmeter could cause a voltage drop so that motor voltage would be lower than that of the line. Similarly, if the potential coils are not disconnected when the current is read, ampere readings are made high by the extra current passing through these two resistances. (b) Now the wattmeter reads its own losses plus those of the ammeter. The ammeter reads the correct motor current but the voltage at the motor is lower than that of the line by the impedance drops in both the ammeter and the current coil of the wattmeter.

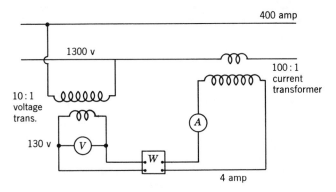

Figure 4-18. Use of instrument transformers in a single phase line. If the wattmeter read 260 w, the real power of the circuit would be 260 × 10 ratio × 100 ratio = 260,000 w. The power factor is 260,000/(13.00 × 400) or 50%.

4-11 Power Measurements—Instrument Transformers

Mention has been made of the fact that the coils of these meters cannot carry a very large current. Yet high-voltage heavy-current lines must be metered. This is done through instrument transformers; for instance, a 100:1 "current transformer" might be provided which steps down the current in exactly that ratio. For accuracy, the secondary and primary current must show no phase shift. Similarly, a small, well-insulated transformer can be obtained for stepping down the voltage in such accurate ratios as 10:1, 100:1, etc. These are then used in the metering circuits, as shown in Figure 4-18.

4-12 Power Measurements—Three-Phase Systems

If the nature of a three-phase load is such that we can connect meters into each leg of a delta load or to the neutral of a Y-connected load, then the obvious method of making measurements is to obtain amperes, volts, and watts in each leg and add the watts of each to obtain total power. Such a scheme with a Y-connected load is shown in Figure 4-19.

This is not always possible. The load to be measured may be that of an entire factory or of a three-lead motor which is built for three-phase operation but with internal connections completely unknown. The total power in such a circuit, regardless of its power factor or degree of unbalance, can be read by two wattmeters connected only in the

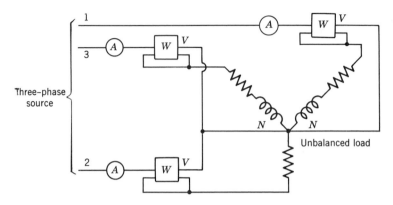

Figure 4-19. Measurement of power, volts, and ampere is an unbalanced Y-connected load by three sets of meters; possible only when the neutral is available for connections. The voltage connection is on the side of the wattmeter marked V.

external supply circuit. This is the universally used method[3] (see Figure 4-20).

Even with a balanced load the two wattmeters do not always read the same. At a load power factor of 50%, one wattmeter reads zero and the other reads correctly the entire circuit power. At power factors below 50%, one meter reads negatively and its results must be subtracted from the watts read on the other meter. The phasor diagrams and mathematics proving these relationships are beyond the scope of this book.

[3] While experimenting with the T-connected transformers in 1894 at Cornell, J. D. E. Duncan used two wattmeters on the two-phase circuit and then reconnected them on the three-phase side. Because he measured about the same amount of power in each case, he went on to prove by phasor diagrams that the two-wattmeter method was correct for three phases. He described this in *Electrical World*.

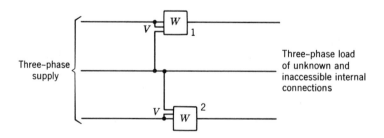

Figure 4-20. Connections for the two-wattmeter method of reading total three-phase power regardless of power factor or unbalance. $W_1 + W_2 = \sqrt{3}VI \times$ power factor.

4-13   Power Distribution

Although this subject may seem unrelated to motor application, it should be pointed out that power distribution in every industrialized country is dependent on the transformer. Without it, the rural dweller would not have electric power (unless he operated his own generating plant) and all industry would have to be clustered around the power plant. This is because, without the transformer, power would have to be generated and utilized at the same voltage.

Suppose we examine a case in which 12 kva were to be transmitted a distance of 1000 ft at a voltage of 120. Refer to Figure 4-21 which shows the essential data.

A No. 6 wire is chosen, which has a resistance of 0.395 Ω/1000 ft. The two wires would weigh 160 lb.

Resistance of both lines is 2 $\times$ 0.395 or 0.79 Ω.

The *IR* drop in both lines with 100 amp is

$$100 \times 0.79 = 79 \text{ v.}$$

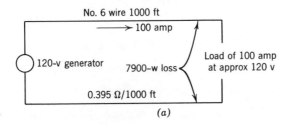

(a)

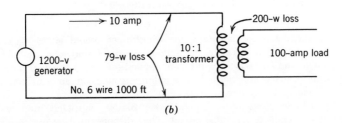

(b)

Figure 4-21. Contrasting the losses in attempting to transmit power at low and high voltages. [Actually such lines display reactance drop (*IX*) as well as *IR* drop.] (a) Attempting to transmit 100 amp at this voltage results in excessive drop and loss unless the wire size is increased many times. (b) The same load is now applied by a high voltage alternator and a transformer at the load center to give the desired voltage. Only a small loss results.

Volts at receiving end

$$120 - 79 \text{ or } 41 \text{ v.}$$

Obviously this distribution line is useless.

Loss in lines,

$$I^2R = 10,000 \times 0.79 \text{ or } 7900 \text{ w.}$$

Over half of the power would be lost in the line.

Now consider the case of Figure 4-21b. Here the power is generated at 1200 v and then stepped down at the load end to 120 v by a 10:1 ratio transformer.

$$I_{\text{line}} = 10 \text{ amp.}$$

IR drop in both lines,

$$10 \times 0.79 \text{ or } 7.9 \text{ v.}$$

$$V_{\text{transformer primary}} = 1192.1 \text{ v.}$$

$$V_{\text{load}} = \frac{1192.1}{10} \text{ or } 119.2 \text{ v.}$$

Loss in lines,

$$I^2R = 100 \times 0.79 \text{ or } 79 \text{ w.}$$

$$\text{probable loss in transformer} = 200 \text{ w.}$$

Load voltage and overall efficiency are satisfactory.

Of course, this is an imaginary and exaggerated case, but it is selected to illustrate the need for transmitting large amounts of power at high voltage to reduce the current, weight, and cost of the conductors. It has all been made possible by the transformer.[4]

[4] Probably the credit for inventing the transformer should go to William Stanley. However, inventions are rarely the original work of one man. In 1882 Gaulard, a Frenchman, and Gibbs, an Englishman, applied for a British patent for an induction coil having primary and secondary windings wound on an iron core which did not form a complete magnetic circuit.

In 1884 three Hungarians, Zipernowsky, Blathy, and Deri, improved on this, using a more complete magnetic circuit. They demonstrated its practicality by using it for lighting electric lamps at the Hungarian National Exposition. They coined the word "transformer."

In 1885 George Westinghouse obtained the American patent rights for the Gaulard and Gibbs patent.

All of the early workers in this field had been disturbed by the necessity of using high-resistance wire in the primary to reduce the current.

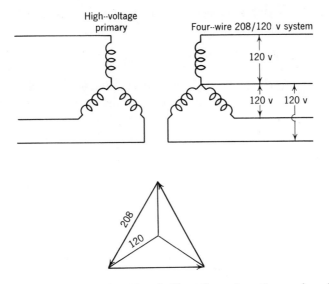

High--voltage primary

Four--wire 208/120 v system

120 v

120 v   120 v

208

120

Figure 4-22. Transformers built with suitable ratios reduce the supply voltage to a four wire Y-connected system with 120 v on each leg and 208 v between the ends of the Y.

4-14 The 208-v Network

Figure 4-22 shows a four-wire distribution system useful for both three-phase motor loads and single-phase lighting or similar loads. If these four wires are distributed through an area, or through an office building, or suburban center, the number of transformers and total conductor area can be reduced, which cannot be done with separate single-phase and polyphase distribution systems.

This network is described here because of the difficulty it presents to motor manufacturers and sometimes to motor users. A three-phase motor is supposed to have minimum values of starting and maximum torques (turning effort) guaranteed and set by national standards. These reduce as the square of the voltage. A standard three-phase motor, designed for 220 or 230 v, applied to such a system may have its torques

It was William Stanley, an employee of Westinghouse, who conceived the idea that it was an induced emf in the primary coil that was most effective in reducing the current and that only a few resistive ohms were needed in the winding. The iron core should make a complete and low-reluctance path for the flux. In that same year (1885) he arranged to generate 500 ac, transmitted it for about 1 mile, and stepped it down to 100 for lighting. It operated at 133 cycles.

reduced by 11 or 18.5% when operating on 208 v. This may be trouble-some. The motor manufacturer then has two choices:

1. He can design his standard line of three-phase motors with enough torque so that even at 208 v they meet the motor industry standards. This has some drawbacks, e.g., excessive starting current when operated at standard voltage or increased amount of active material in the motor.

2. He can design and stock a line of 208-v motors and another line of 220- or 230-v motors. This could tie up millions of dollars in inventory.

The 208-v system is not looked upon as a blessing by the motor industry.

4-15   The Power-System Battle

In spite of the presence of the 208-v system, in this country we are favored by a standardization of frequency and voltage that is taken for granted.

Here, a householder can move from one state to another, certain that all of his electrical appliances will still be workable. A factory also can move equipment from one plant to another anywhere in the United States.

In other countries various voltages and frequencies appear in different areas. It was not until after World War II that eastern Canada changed from 25 to 60 cycles. Upper New York state had 25 cycles for many years. Parts of Michigan and California were supplied by 50-cycle power until about 30 years ago.

The U.S. Department of Commerce publishes a book showing standard voltages and frequencies throughout the world. Any motor builder who supplies motors to equipment manufacturers who do a worldwide business has to design motors for operation on 15, 25, 30, 40, 50, 60, and $66\frac{2}{3}$ cycles as well as for a variety of voltages.[5]

Standardization on 60 cycles, 115/230 v, for single-phase motors and 230/460 v[6] for industry was not accomplished in this country without bitter words and vicious quarrels.

The first battle occurred when it was decided to use some of the hydraulic power of Niagara Falls for generating electricity. George Westinghouse, with the rights to the transformer patents as well as to

---

[5] A few years ago one business-machine manufacturer presented the Pope with a gold-plated electric typewriter. The motor had to be designed for $66\frac{2}{3}$ cycles.

[6] Raised in 1965 from 220/440 with some reluctance by the motor industry, after pressure from the Power Company Association for many years.

those of Shallenberger for the kilowatt-hour meter, and with experience with a few ac distribution lines, led the group favoring ac generation.

Edison's pioneer central station on Pearl Street in New York had been operating since 1882, supplying lighting service over a radius of about ¾ mile. He favored direct current and wrote in the North American Review: "There is no plea which will justify the use of high tension and alternating current in either a scientific or commercial sense—my personal desire would be to prohibit entirely the use of alternating currents . . ."

Lord Kelvin (formerly Professor Sir William Thomson) as president of the International Niagara Commission favored direct current and cabled his other technical advisors: "Trust you avoid gigantic mistake of adopting alternating current."

Editorials appeared in leading newspapers condemning the "deadly alternating current." Thus the argument continued until finally George Westinghouse and his group won, and by 1895 the Niagara Falls plant was in operation with an output of three-phase 25-cycle generators.[7]

Nevertheless, power stations in those days operated separate generators for arc-lighting systems, 500-v generators for street railways, dc dynamos for lighting nearby areas, and single-phase generators for lighting remote districts. Standardization was a slow process.

4-16   The "Deadly Alternating Current"

The electric chair, it must be admitted, has nothing to do with motors. But it is of interest to note that alternating current has been labeled from the start as "deadly" by proponents of the Edison dc system and newspaper editors. If so, the next logical step was to make use of this quality.

And so, in Frank Leslie's *Illustrated Newspaper*, June 8, 1889, we find a drawing of a man stretched out apparently at ease (his face is covered by a black mask, but his position looks comfortable) with the sheriff ready to throw the switch. The account reads, in part:

[7] Professor H. A. Rowland of Johns Hopkins had been hired as a sort of referee and consultant by the Commission. He helped influence the decision for alternating current, but after his fee was ignored for several years, he took the case to court. While on the witness stand, he was asked by a lawyer for the Commission as to whom he considered to be the outstanding electrical physicist in the world. "I am," said Rowland. Later he was chided by his associates for his seeming lack of modesty. "Ah, but I was under oath," he replied.

"The necessity of a more humane and certain means of execution than hanging was illustrated by the sickening details of a recent triple hanging in Missouri, where the stretching of two ropes and the breaking of a third, slowly tortured the three criminals to death. William Kemmler, who brutally murdered his paramour at Buffalo, N.Y. need fear no such bungling work, since he is the first to be convicted under the new law, and will be executed by Westinghouse alternating current . . . ."

There is no evidence that the well-known slogan, "You can be *sure* if it's Westinghouse," originated at this time.

Chapter 5

The Polyphase Induction Motor, Operation and Types

We start this chapter by showing the reader how a two-phase or three-phase supply connected to suitable windings in the stator of an induction motor produces a rotating flux in the air gap (see Figure 5-1); a veritable whirlpool of flux which rotates at a speed fixed by the number of poles in which the winding is arranged and by the frequency of the supply.

A two-phase motor is used first to illustrate this phenomenon, not because such motors are of much commercial importance but because the principles can be more readily grasped if only two phases are considered.

Refer to Figure 5-2. This represents a two-phase stator winding of two poles. It may appear to have four poles, but bear in mind that coil sides of both phases 1 and 2 are combined as one pole. Note stator a. The top conductors have current coming out of the page, the bottom sides of the same coil show current going into the page. By the right-hand rule this would build up a flux system going across the air gap to the right. This represents point a on the two sine waves of current. Phase 2 has zero current at this instant.

Now refer to instant b on the current waves. Both phase 1 and phase 2 have a lesser current, but still going in the same direction as at point a. The magnetizing forces of both phase windings combine to give a flux now pointing 45° below the horizontal.

Refer to stator (c) and note that the current in phase 1 is now zero with a maximum current in phase 2. Phase-2 coils now build up a magnetizing force and a flux pointing vertically downward. The distance from (a) to (b) to (c) represents one-fourth of a cycle. The flux in the air gap has moved one-quarter of a revolution. By continuing this step-by-step process, we would find that in one cycle (360°) the flux also

Figure 5-1. Cutaway view of small polyphase motor. (Courtesy of Robbins & Myers, Inc.)

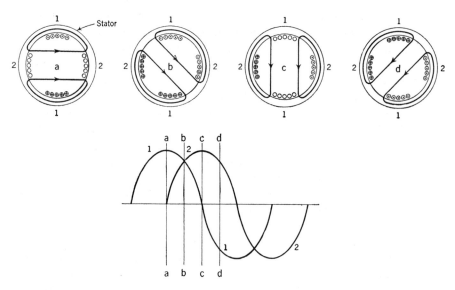

Figure 5-2. A two-pole, two-phase winding connected to a suitable supply builds up a magnetizing force or flux that rotates as the current supply goes through its cycles.

rotated 360° in the air gap. If the supply is 60 cycles, then the flux rotates one revolution in $\frac{1}{60}$ sec. This is 3600 rpm and is called the *synchronous speed*.

We found in the example of Chapter 1 that many more ampere-turns are needed to set up flux through air than through iron or steel. Therefore these stators as shown require excessive current until we put a steel rotor inside the stator with clearance between the two of, say, 0.008 to 0.100 in., depending on motor size. There are about half a dozen different rotor constructions that can be put on a shaft and used inside such a stator. The construction used determines the type of motor and the characteristics desired.[1] Reverse the leads of any one phase and the flux reverses.

5-1   Three-Phase Stator

Note Figure 5-3 in which three sets of coils are distributed around the periphery of the stator. Sides of the same coil are shown 180° opposite. These coils are 120° apart. Examine the individual points in time, (*a*), (*b*) (*c*), and (*d*), noting the instantaneous values of current and the position of the magnet motive force (mmf) and flux that results. Once again, in one-half cycle of current, we obtain 180° of flux rotation. If 60 cycles were applied, one revolution of the flux wave would occur in $\frac{1}{60}$ sec, or the flux revolves at 3600 rpm.

Now suppose we remove these windings, each of which spans 180°, and replace them with coils which span 90°. In other words, these three coils are now spread over only one-half of the periphery of the stator. A similar set of coils are placed in the other half. This becomes a four-pole winding. As the current goes through its 360° of a cycle, it moves the flux over one pair of poles only. Now with 60 cycles supplied, the flux moves one-half revolution per cycle and, hence, the speed is 1,800 rpm. This is the synchronous speed.

Only in a two-pole machine does the flux cross the entire diameter of the inside of the stator. In a four-pole motor the paths are as shown in Figure 5-4.

[1] A dramatic means of displaying the presence of this magnetic rotating field is to take a bare stator with windings but no end enclosures, reduce the voltage so the windings do not overheat, and put a short piece of steel pipe (or even a can of tomatoes) at the bottom of the stator. It races around the inside surface in an impressive manner.

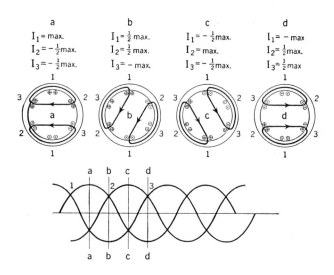

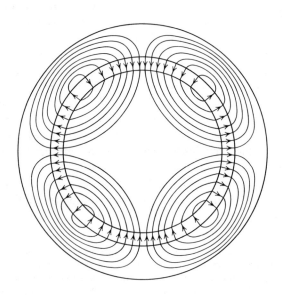

Figure 5-3. In this two-pole, three-phase stator the currents are shown at various points in time. The resulting magnetomotive force of the three sets of coils is shown to rotate.

Figure 5-4. In a four-pole motor the flux distribution is arrested in space at a given instant.

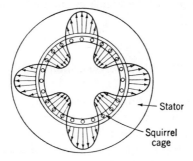

Figure 5-5. Although the flux must make a complete circuit, as shown in Figure 5-4, it is also convenient to represent its direction and magnitude by the arrows. The windings, connected to a 60-cycle supply, would cause the flux to rotate at 900 rpm.

5-2 Three-Phase Y-Connected Windings

If a stator has 36 slots and the designer wishes to place a four-pole three-phase winding in it, his first calculation involves dividing the slots by the poles, yielding nine slots (see Figure 5-5). He can then spread any one-phase winding in three slots under each pole. If a six-pole winding were desired, there would then be six slots per pole, and each phase winding would be distributed in two slots under each pole.

The actual windings and connections for the latter case are shown in Figure 5-6. We need only glance at this figure to realize that it

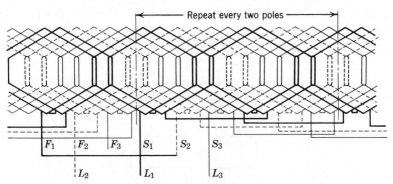

Figure 5-6. If the stator is cut open at one spot and laid out flat, it is more convenient to show a developed winding. There are two slots per phase per pole. This is a lap winding because the coil ends lap over one another. It is also distinguished as a double-layer winding, since there are two coil sides per slot. It is a Y-connected single circuit. Reversing any two lines to the supply reverses the rotation of the motor.

presents a field best left in the hands of the motor designer or the motor repair shop personnel.

5-3   Rotors: The Simple Synchronous Motor

Now that we have a rotating magnetic field, what form will the rotor take by which we can get motor action? Perhaps the simplest rotor, most easily understood, is a bar of iron or steel mounted on a shaft and having salient (projecting) poles equal to the number of poles wound on the stator (see Figure 5-7). If we could get such a rotor up to speed, close to the synchronous speed of the four-pole field, it "would pull into synchronism." The rotating field would select the paths of minimum reluctance, which would be through the poles, and carry the rotor around with it. This rotor would then turn at exactly 1800 rpm with a sixty-cycle supply. It is a synchronous motor. (The word comes from the Greek, meaning "in rhythm.") These motor types are covered in Chapter 9.

5-4   Squirrel-Cage Rotors

A lamination for a squirrel-cage rotor is shown in Figure 5-8. This is usually punched from the center of the stator and, hence, is a special steel having more or less silicon in it, depending upon the grade and cost as selected for the stator.

Such a lamination will be built up into stacks of various lengths, depending upon the design. The one shown in Figure 5-8 might have rectangular copper bars in each slot. Other rotor laminations might have round slots for round bars. These bars may extend anywhere from ¼

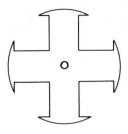

Figure 5-7. This four-pole structure, built of laminated steel or even of cast iron, mounted on a shaft in a four-pole excited stator rotates at synchronous speed, tying in with the rotating flux. This is basically a synchronous motor.

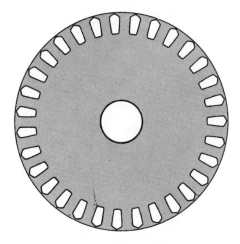

Figure 5-8. Reproduction of a rotor lamination. Copper bars are placed in the slots of a stack of these laminations to form a part of a simple rotor winding.

to 1 in., or more, at each end of the stack. A circular ring of copper is then welded, brazed, or soldered to each bar, so that the electric circuit apparently seems to be short-circuited in the form shown in Figure 5-9.

Place this rotor of laminations, copper bars embedded in the slots, and end rings electrically connected to each bar in a suitable stator. Exciting the stator immediately builds up a flux system which crosses the air gap, goes through the teeth of the rotor, and completes its circuit through the core of the rotor under the teeth. Refer again to Figure 5-4.

The flux rotates at synchronous speed and, as it does, cuts the rotor

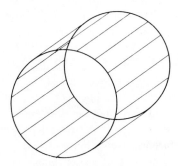

Figure 5-9. Basically, the rotor winding consists of this type of electric circuit.

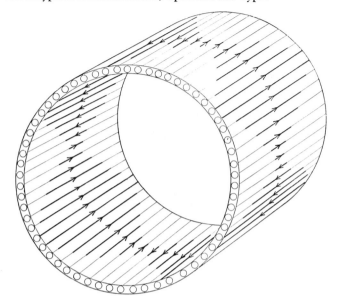

Figure 5-10. The actual directions of the voltages generated are indicated by arrows. The magnitude of the voltages is given by the length of the arrows. The schematic representation of the rotor squirrel-cage winding contains four paths for the voltages and currents, since it is assumed to be a four-pole stator.

bars, inducing a voltage in each. The magnitude of the voltage in any one bar at any instant depends on the magnetic density of the field at that point. (It is assumed that the flux density from point to point around the air gap varies as a sine wave.) Hence the voltage in the bars will take on magnitude and directions, as shown in Figure 5-10. These voltages cause currents to flow through the bars in one direction, then through an end ring, back through adjacent bars and the opposite end ring, so that a complete electric circuit is made for each pole. Four such sets of circuits and circulating currents are built up if the rotor is placed in a four-pole field. (The same rotor, incidently, placed in a six- or eight-pole field would change the number of circuits automatically and might work equally well with all of the stators.)

Because the bars are embedded in steel, they have a high reactance. The currents lag the voltages and, as the end rings are fed currents of various magnitudes from each bar, the end-ring current is not uniform throughout its circumference (see Figure 5-11). When the rotor is stationary, the flux rotates past it at synchronous speed, so that the rotor currents and voltages are of the same frequency as the supply.

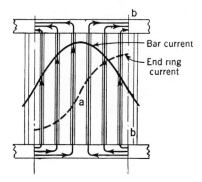

Bar current
End ring current

Figure 5-11. The currents differ in each bar under one pole and the distribution of current in each end ring is not a constant at all points.

5-5 Motor Action

In all of this detail we must not overlook the fundamental fact that we have conductors (rotor bars) carrying current in a magnetic field. The field built up around each bar reacts with the air-gap flux, exerting a force on each bar which causes the rotor to rotate. This is similar to the principle illustrated in Figure 1-14, except that now the field flux is rotating also.

Another rather loose way of explaining this action is to say that the currents in the bars build up poles (equal in number to those of the stator), and these poles are attracted by the rotating poles built up by the stator winding and follow after it. Regardless of the point of view, the motor starts and runs. The force exerted on each bar depends on the strength of the magnetic field in the air gap and the current in each bar. The total forces add up to the turning effort or torque of the motor.

In every motor we have a combination of motor action and generation action. Recall that the magnitude of the voltage generated in a conductor depends on its rate of cutting flux (10^8 lines cut per second give 1 v). Hence, when the rotor is stationary, the flux is cutting the rotor bars at the maximum (synchronous) speed. The highest rotor voltage occurs then and, granting that the resistance and reactance of the rotor circuits stay constant (the causes of slight variation in these values will be discussed later), the rotor current will also be a maximum.

Now the rotor accelerates. Consider that it has gotten up to two-thirds of synchronous speed, so that, in a four-pole motor, it is momentarily running at 1200 rpm. At this instant, with the stator flux revolving at 1800 rpm and the rotor bars at 1200 rpm, the *relative* speed between them is only 600 rpm. The rate of flux cutting is one-third that of standstill, the voltage in the bars is one-third that at standstill, and the frequency of the voltage and current in the rotor is now only 20 cycles.

The motor continues to accelerate until (let us assume) it reaches a speed of 1798 rpm. Now the voltage generated in the cage is 2/1800 of the value at starting, the frequency of the rotor voltage and current is 2/1800 of 60 cycles. The motor may continue to run constantly at 1798 rpm after going through its acceleration and with no load connected to its shaft.

Why does it not continue to accelerate to synchronous speed? If it did, the rotor bars would rotate at exactly the same speed as the air-gap flux. The bars would not be cutting the flux, no voltage would be generated in them, and no current would flow in the bars to build up a magnetic field which could react with the stator field to produce a magnetic force on them. The motor would develop no torque or turning effort until it slowed down slightly so that the air-gap flux again would cut past the rotor. Hence an induction motor always runs at slightly less than synchronous speed, even with no load connected to the shaft. The final speed of 1798 rpm assumed here is the result of bearing friction and the power to drive the fan that is usually connected internally to the shaft. The rotor is always loaded internally to that extent.

The difference between synchronous rpm and rotor rpm is called the *slip*. In this case the slip is 2 rpm. It is also expressed as

$$\text{slip, \%} = \frac{\text{synchronous rpm} - \text{actual rpm}}{\text{synchronous rpm}} \times 100 \qquad (5\text{-}1)$$

5-6   Internal Reactions when A Load Is Applied

Suppose now that a mechanical load is applied to the shaft, say, a fan or a pump. This exerts a retarding torque so that the motor immediately slows down. As it does, the *relative* speed between the synchronously rotating air-gap flux and the rotor bars increases. More voltage is generated in the bars and more current flows to react with the flux to produce more turning effort. The motor continues to slow down until this internal torque developed exactly equals the torque required to turn the load. Usually this reduction in speed at rated load on a motor gives a slip of 5% or less.

Thus, at full load, six-, four-, and two-pole motors may rotate thus:

1200 − 5% of 1200 or 1140 rpm (nominal rated speed).
1800 − 5% of 1800 or 1710 rpm (1725 rpm is more common).
3600 − 5% of 3600 or 3420 rpm (3450 rpm is more common).

If a fifty-cycle supply were used, the synchronous speeds would be 1000, 1500, and 3000 rpm respectively.

In general,

$$\text{Synchronous rpm} = \frac{\text{frequency} \times 120}{\text{poles}}. \qquad (5\text{-}2)$$

As the load current flows in the rotor bars and end rings, the magnetizing force it builds up opposes the air-gap flux built up by the stator winding, just as does the secondary current in a transformer (see Chapter 4). The primary or stator current must maintain a high enough value to neutralize the opposing magnetizing force of the rotor and still provide the flux necessary to give the stator a counter electromotive force. In this way the induction motor is exactly like a transformer. At no load the input current is largely that required for magnetizing the steel in the stator and rotor paths and driving the flux across the air gap. As such it lags the applied voltage by 90°. However, even running near synchronous speed, at no load there are some losses; hence the real no-load current lags less than 90°, having a slight in-phase component representing these losses.

One might say that in an induction motor the slip is a sort of "governor" which controls the torque developed and the current that the motor draws from the line.

5-7   Work, Torque, and Power

Work is the operation of overcoming an opposing force through a certain distance. It is measured in foot-pounds (ft-lb). Thus a man weighing 150 lb, climbing a stair 10 ft high, does 1500 ft-lb of work. He would do the same amount of work if he raised a 10-ft weight from the bottom of a 150-ft well.

Power is the rate of doing work. If the man took 20 sec to walk up the stairs but later ran up in 10 sec, he would be doing the same amount of work, but in the second case he would be doing it in half of the time or at twice the *rate* and, therefore, putting out twice as much power. Hence, power involves both work and time.

By definition, as probably erroneously decided by our forefathers,

$$1 \text{ hp} = 33{,}000 \text{ ft-lb/min} \qquad (5\text{-}3)$$
$$= 550 \text{ ft-lb/sec} \qquad (5\text{-}4)$$
$$= 746 \text{ w.}$$

Torque has been mentioned as the tendency to rotate. It can exist whether there is rotation or not.

Figure 5-12 shows a sketch of a man with a pipe wrench attempting to loosen a pipe from a coupling. Pulling with a force of 60 lb at the radius or 1 ft, he is unable to loosen the pipe. With a larger pipe wrench with a longer handle of 2 ft and with the same pull of 60 lb, he can turn the pipe. In the first case he exerted a torque (attempted rotating

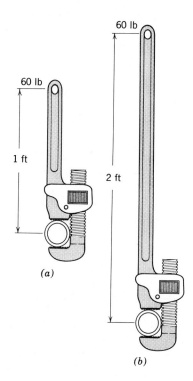

Figure 5-12. An illustration of force in a circular path called *torque*. (a) A 60-pound pull on this wrench fails to loosen the pipe. This effort is called torque because, had it moved, it would have been a force exerted in a circular path. (b) The neighbor's wrench with a longer handle turns the pipe. The torque exerted involves both length of arm and force. In this case the torque is 2 × 60 or 120 lb-ft. Had it required this same torque throughout the entire turn, the *work* done would have involved distance. This would be the circumference of a 4-ft circle.

motion) of 60 lb $\times$ 1 ft or 60 lb-ft. In the second case he exerted a torque of 60 lb $\times$ 2 ft or 120 lb-ft. Hence torque involves not only pull or push but also the radius at which the force is applied. Note, too, that the units are pound-feet, or just the opposite of the term foot-pounds for *work*.

Work involves *distance*. If a man with the 2-ft pipe wrench had to exert 60-lb pull, the entire *distance* is the diameter of the circle of 2-ft radius. Hence,

$$2\pi \times R = \text{the circumference or } 4\pi \text{ ft,}$$

$$\text{work done} = 4\pi \text{ ft} \times 60 \text{ lb or } 754 \text{ ft-lb.}$$

If it takes him 30 sec to turn the pipe, he would be rotating it at $\frac{1}{2}$ rpm. Then,

$$\frac{754 \times \frac{1}{2}}{33,000} = 0.01135 \text{ hp.}$$

We can generalize this by writing,

$$\text{hp} = \frac{\text{pull} \times \text{length of arm} \times 2\pi \times \text{rpm}}{33,000}$$

$$= \frac{\text{torque, lb-ft} \times 2\pi \text{ rpm}}{33,000}$$

$$= \frac{\text{torque, lb-ft} \times \text{rpm}}{5252}. \qquad (5\text{-}5)$$

What horsepower is developed if an electric motor operating at 1725 rpm is turning a load that requires 3.05 lb-ft of torque?

$$\text{hp} = \frac{3.05 \times 1725}{5252} \text{ or } 1.0.$$

Here is a fact that is very significant. Every motor of any speed must develop a definite torque to justify its rated horsepower. Thus, a $1\frac{1}{2}$-hp four-pole motor must have a full-load torque of 4.57 lb-ft. A 1-hp 3450-rpm motor must have a full-load torque of 1.52 lb-ft, and so on.

In general,

$$\text{full-load torque} = \frac{\text{hp} \times 5252}{\text{rpm}} \text{ lb-ft.} \qquad (5\text{-}6)$$

It is not unlikely that a prospective motor buyer may tell a motor manufacturer that he needs a 1-hp 1725-rpm motor for a load requiring

4 lb-ft of torque. The obvious answer is that a 1-hp motor will do this, but it will be overloaded about 30% and may overheat if used continuously. If used intermittently, with a chance to cool off between operating periods, it may be satisfactory. As an alternative, a $1\frac{1}{2}$-hp motor has a full-load torque of 4.57 lb-ft and would be slightly under-loaded, on application, but would operate cooler.

The fact that torque was definitely linked to horsepower may have been vague (if not totally lacking) in the prospective buyer's mind.

5-8   Losses and Efficiency

When a polyphase motor is operating, all of the power put into it (in watts) does not appear as useful output in the shaft. To do so it would have to be 100% efficient, with no losses. The only electric equipment which is 100% efficient is an electric heater. All of the energy put in appears as heat [1 kwhr = 3413 British Thermal Units (Btu)].

But an electric motor has four main sources of loss, all of which appear as heat.

1. Core or iron loss, depending chiefly on the flux density, the frequency, and the grade of steel used.

2. Copper or I^2R loss in the stator winding.

3. I^2R loss in the rotor cage.

4. Friction in the bearings and windage loss from the rotation of the rotor including the fan, if used.

Core Loss. To explain this let us return to the transformer for a simple yet applicable illustration. Refer to Figure 5-13. Here we see the effect of solid iron or steel used as a magnetic core. To reduce the heat loss caused by these induced eddy currents, the core is built up of thin layers of steel, each one more or less insulated from the other.

A transformer was used to illustrate this idea only because of the apparent simplicity of the circuit. In an induction motor the teeth and the yoke are all subjected to varying fluxes which would give the same effect, but the physical relationships are more difficult to visualize.

Most motors are built of steel laminations of No. 24 gage 0.025 in. thick. As the contours and slots are punched out by automatic presses, a "work-hardening" takes place which reduces the magnetic qualities. Laminations are then heat-treated to about 1500°F in a controlled atmosphere which anneals them and reduces the above effect. It also puts an iron oxide on the surfaces, which is a good insulator. Thus one lamination is insulated from another. (Or lamination steel can be purchased which already has a thin varnish coat on each side for the same purpose.)

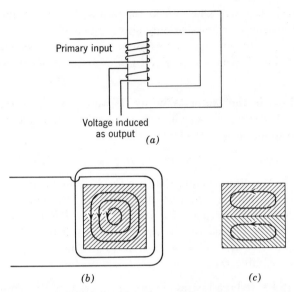

Figure 5-13. A representation of eddy current losses set up in the magnetic core and how they can be reduced. (*a*) A transformer with a solid iron core and the primary and secondary windings on the same leg. (*b*) An enlarged cross section of the solid iron core showing the circulating or eddy currents induced in the core. Just as a voltage is induced in the primary as a counterelectromotive force, or in the secondary as output, so does the solid iron core represent one short-circuited turn. Under certain conditions these eddy currents could make the core red hot. (*c*) The iron core is now made up of two pieces, insulated from each other. The flux to induce voltage in each piece is one-half the total. The cross section of the iron as a conductor is one-half, doubling the resistance. As a result the eddy current losses are now one-quarter of the original value.

Actually, motors are cursed with two kinds of eddy-current losses; those that occur within each lamination and those that flow from one or more laminations into others as a result of the surface insulation being inadequate. The latter occurs occasionally in production and results in motors with unexpected high losses.

Hysteresis loss is the other component of core loss. The easiest way to explain it is to point out that, when a section of steel or iron is magnetized in one direction, its molecules all line up in one manner. Reversing the magnetism causes the molecules to turn over. The internal friction loss is called *hysteresis*. It will vary directly with the frequency. In a motor with a magnetic field rotating so that first one area is a north pole and then an instant later is a south pole, it can be seen that hysteresis loss will show up in the stator, but would be very small in a rotor because the rotor sees only slip frequency.

Engineers usually employ curves of total core loss per cubic inch of steel versus flux density for a given grade and frequency to determine core loss.

Copper Loss in the Stator Winding. If we can measure the resistance of each leg of a three-phase winding and read the average current input per line, the copper loss of the stator is, then,

$$3 \times I^2_{av} \times R_{1,av} \text{ w.} \tag{5-7}$$

I²R Loss in the Rotor Cage. Because the current in the rotor bars and end rings is unevenly distributed, calculation of the loss is complex but possible. The easiest method of determining the I^2R loss in the rotor is by test. It can be proved that if the slip at any load is 5% then 5% of the power put into the rotor is the I^2R loss. This will be illustrated in the following sample test.

Friction and Windage Losses. These losses are usually estimated unless test data are available.

The term *efficiency* is often used loosely, but to the engineer it means

$$\text{efficiency} = \frac{\text{output}}{\text{input}} \tag{5-8}$$

$$= \frac{\text{input} - \text{losses}}{\text{input}}. \tag{5-9}$$

5-9   Motor Tests

To test a three-phase motor requires a laboratory equipped with a source of three phase power which can be varied over a wide range of voltage*; a similar source of power on which the voltage can be held constant; wattmeters, ammeters, and voltmeters of suitable capacities; a means of reading speed very accurately; a prony brake for load tests* and a dynamometer capable of carrying desired rated motor loads continuously. (Those items marked with an asterisk could be eliminated.)

1. The tests usually made are "running-performance tests" from no load to 125% of load.

2. Speed-torque tests showing how the developed torque varied with speed.

3. Heat runs indicating the final operating temperature of the motor windings (and other components) when operating continuously at a stated load.

Occasionally we may make a no-load "saturation" run, which enables the engineer to calculate the separate losses in the motor. This is the test that requires the variable voltage source.

Tests 1 and 3 require a means of loading the motor *so that its mechanical output can be measured accurately.*

The Prony Brake. If we put a drum on the motor shaft and fasten a brake band around this drum so that the breaking action is transmitted through a lever arm, we have a means of loading a motor. The mechanical

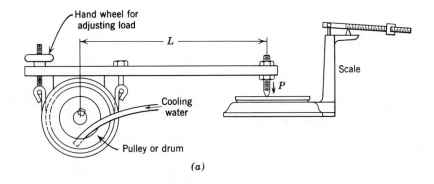

(a)

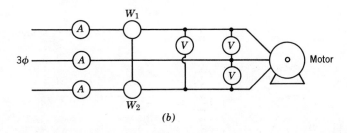

(b)

Figure 5-14. The above set-up (adding a means of accurate speed reading) enables the test group to determine watts input, amperes input, power factor, speed, and efficiency of the motor for a range of horsepower. Unfortunately the brake occasionally grabs, the platform scale vibrates, and the operator adjusting the brake would do well to wear a raincoat. (*a*) A prony brake connected to a motor shaft for reading mechanical output. Torque = PL: hp output = (PL × rpm)/5252 (subtract the weight of the arm from *P*). (*b*) The motor input in watts, amperes, and volts is read at various outputs, as determined by the tightness of the brake drum.

output is all absorbed as heat in the brake and drum. But if we place
the end of the lever arm on a scale, the torque transmitted to the load
can be read, say in pounds on the scale. This times the lever arm in
feet gives the pound-feet of torque. Read the speed accurately and, from
Eq. 5-5, we can then calculate the horsepower output of the motor
(see Figure 5-14 showing the usual arrangement).

The Dynamometer. Fortunately, the prony brake is obsolete in most
laboratories. Its occasional lack of stability made the results uncertain.
As a loading device it has been replaced mainly by the dynamometer,
which can be a very sophisticated piece of equipment, with electronic
pickups for reading speed to $\frac{1}{2}$ rpm and other attachments which will
be discussed later. We shall describe the most simple type to illustrate
the principle.

Imagine a dc generator with no base or feet. It has the usual bearing
inside the hub and on the shaft. Suppose, now, that we add another
set of large ball bearings on the *outside* of the hub which fit into recesses
in pedestals mounted on a base. The body of the dc generator could
spin around. Instead, we add an arm of definite length radially to the
body. This arm is attached to a scale. Now the body can turn only
within the limits of the spring inside the scale (see Figure 5-15). Flexible
leads bring the output of this generator to large resistors, the function
of which is merely to absorb the the dc output. We now have a dynamometer. Couple it to a motor and, by changing the resistance load
on the generator, the motor load can be varied. But as action is equal
to reaction, the torque put out by the motor can be measured on the

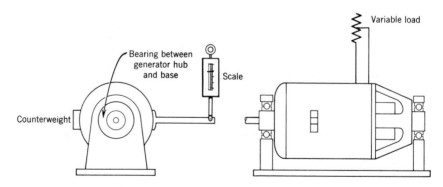

Figure 5-15. Representation of a dc generator used as a dynamometer. The generator not only loads the motor under test but indicates the motor torque output
by the lever arm and scale.

scale which is deflected in the opposite direction from that of the motor rotation. The *generator output* need not be measured. Again the horsepower output at any load is (power loss × rpm)/5252.

5-10 Calculating the Performance at Various Loads

This process involves nothing distinctly different from the calculations to which the reader has already been exposed. We shall consider one point of load and carry through an example.

Laboratory data: A ¾-hp motor is to be tested on a dynamometer. The arm of the dynamometer, which is 1½ ft long, is counterbalanced so that scale reading need not be corrected for dead weight of the arm.

Scale reading, 2.66 lb.
Speed, 1658 rpm. (4-pole 60-cycle motor).
Sum of both wattmeter readings, 1298.
Average of three ammeter readings, 1.95.
Applied volts, three phases, 440-v.

Solution:

$$\text{hp output} = \frac{2.66 \times 1.5 \times 1658}{5252}$$

$$= 1.26 \text{ hp (this is an overload).}$$

$$\text{output, w} = 1.26 \times 746 \text{ or } 940.$$

$$\text{efficiency} = \frac{\text{output}}{\text{input}} = \frac{940}{1{,}298} \text{ or } 0.725.$$

$$\text{pf} = \frac{\text{watts}}{\sqrt{3} \text{ volts} \times \text{amp}} = \frac{1{,}298}{\sqrt{3} \times 440 \times 1.95} \text{ or } 0.87.$$

We know that the input amperes average 1.95 (sometimes the currents do not quite balance) and that the input watts are 1298, so we have all of the data necessary to draw at least one series of points on the performance curves as shown in Figures 5-16 and 5-17. These motors do not include the one calculated in the foregoing.

On these curves, for 5- and 50-hp motors respectively, note the no-load current. It represents the magnetizing current to send the flux through the lamination steel and across the air gaps. Increasing the gaps would raise the current considerably. As this current is magnetizing the highly inductive winding, it lags the voltage in each phase by almost 90°.

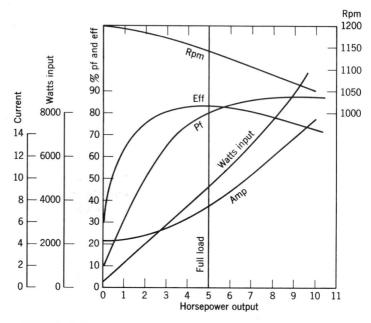

Figure 5-16. Typical performance of a 5 hp, six-pole, three-phase motor tested at 440 v. The full-load torque is 23 lb-ft. Maximum torque is 81 lb-ft and the starting or blocked rotor torque is 70.6 lb-ft. Starting amperes are 45.5.

Hence the low power factor at light loads. The increase in efficiency in the 50-hp motor, 96 versus 83% in the 5-hp rating, is typical of all electric equipment. Higher ratings are naturally more efficient.

The torques shown are the result of further tests to be discussed later.

5-11 Separating the Losses

The reader may never need to be involved in this process of determining the individual losses in a motor, but the process is fairly simple and is useful in showing their relative values.

The saturation curve is obtained by running the motor idle and reading the voltage and watts (and sometimes the current) over a wide range. The voltage is reduced until the speed starts to reduce. Values are plotted as shown in Figure 5-18. I^2R and core losses are practically negligible at low or zero volts. The few watts input remaining must be friction and windage. The motor used in this figure is the $\frac{3}{4}$-hp rating previously used in Section 5-10.

Assume the same input point previously used. We must know the resistance of the winding per leg. It is 14.2 Ω on the 440-v connection, With a current of 1.95, the I^2R loss in the entire stator is

$$3 \times 1.95^2 \times 14.2 \text{ or } 162 \text{ w.}$$

We must now determine the core loss. The idle readings show that the no-load current at rated voltage was 0.95 amp. Therefore the stator copper loss under this condition is

$$3 \times 0.95^2 \times 14.2 \text{ or } 38.3 \text{ w.}$$

The watts at this point were 108.

no-load input − (friction and windage) − (stator I^2R) = core loss.

This assumes that the I^2R loss in the rotor, at no load, is negligible. Therefore,

$$108 - 4 - 38.3 = 65.7 \text{ w (the core loss watts).}$$

In normal operation it is assumed that the motor runs at 440 v and, therefore, the core loss is 65.7 at any load. Furthermore, it nearly all occurs in the stator because the rotor frequency is so low.

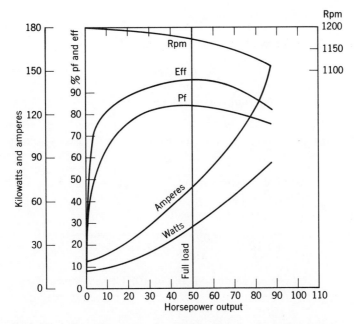

Figure 5-17. Typical performance of a 50 hp, six-pole, three-phase motor tested at 440 v. The full-load torque is 223 lb-ft. The maximum torque is 491 lb-ft. The starting torque and amperes are 330 lb-ft. and 365 amp, respectively. Note the higher full-load speed of this motor.

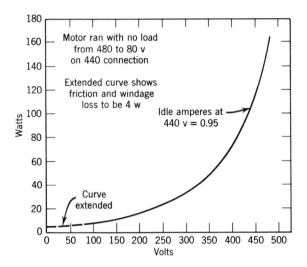

Figure 5-18. A no-load run indicates a friction and windage loss of 4 w. As the curve of watts versus volts is extended to zero volts, the current is so small that I^2R is negligible. The core loss varies as the square of the voltage and is assumed negligible at the extended portion of the curve. The remainder must be friction and windage.

So far we known that, at the load point of 1,298-w input,

$$I^2R_{stator} = 162 \text{ w,}$$
$$\text{core loss} = 65.7 \text{ w,}$$
$$\text{friction and windage} = 4 \text{ w.}$$

It has been mentioned before that of the power which crosses the air gaps into the rotor the same percent is lost in I^2R as the percent of slip.

$$\text{power across gap} = \text{input} - I^2R_{stator} - \text{core loss,}$$
$$1298 \text{ w} - 162 - 65.7 = 1070.3 \text{ w to rotor,}$$
$$\text{speed at this load} = 1658 \text{ rpm,}$$

$$\% \text{ slip} = \frac{1800 - 1658}{1800} \; 100 = 7.9\%,$$

$$I^2R \text{ loss in rotor} = 7.9\% \text{ of } 1070.3 = 88.8 \text{ w,}$$
$$\text{output} = \text{rotor input} - I^2R_{rotor} - \text{friction and windage}$$
$$= 1070.3 - 84.8 - 4 \text{ or } 981.5,$$

$$\text{efficiency} = \frac{\text{output}}{\text{input}} = \frac{981.5}{1,298} \text{ or } 0.756.$$

The dynamometer test showed an efficiency of 0.725 and an output of 940 w vs. 981.5 w by the separate-loss method. This represents a condition that occurs frequently. There are 41.5 w lost somewhere in the motor, accounted for in the actual test but not included in the theoretical calculation of losses. These extra lossses are called *stray load losses*. They are probably caused by wandering currents through the rotor and usually increase with load. Stray load losses may occur in varying degrees in the same line of motors going through production.

5-12   Speed-Torque Curves

If we tighten the brake on the prony brake so that it cannot slip or lock together the armature and body of the dc generator, connecting the coupled motor to the line results only in enough rotation to bring the scale to a balance. In short, the scale reads (using the correct lever arm) the starting or locked rotor torque of the motor. Usually the current is read also. It is five to six times the full-load current, and readings are usually taken quickly because the motor overheats rapidly.

Next the motor is loaded to a number of values as in the full load tests, only now attention is centered on reading the speed, the torque, and possibly the amperes and volts. Loading is continued until we come to a point where the motor becomes noisy and grinds to a stop. Readings just before that point display the *maximum torque* that the motor will develop. This is a very important characteristic. With these readings we can now draw the speed-torque curve for the motor. Refer to Figure 5-19.

In discussing the operational theory of the polyphase induction motor, it was pointed out that at standstill the air-gap flux cut past the rotor bars at the maximum rate and, therefore, the voltage induced in them and the current that flowed through the cage were maximum as compared with any other speed. This shows up on the current curve. But the torque is not a maximum at standstill because the high reactance of the cage makes this current lag too far behind the voltage to be effective. Therefore, we could suspect that, by making the rotor bars of smaller section or reducing the section of the end rings, this higher resistance would give the motor more starting torque. We shall see that this is exactly what happens.

Rotor Resistance Slows Down the Motor. In the running region around synchronous speed, this higher rotor resistance may or may not be a blessing. The motor gets its torque by the reaction of the flux around

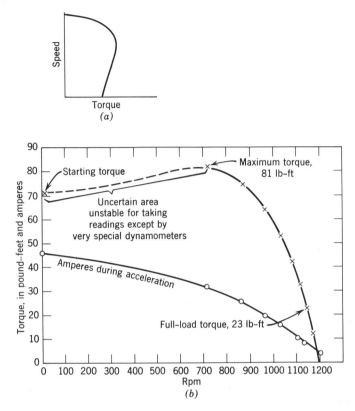

Figure 5-19. Speed-torque curve for the 5 rps motor of Figure 5-15: max torque = 81/23 or 351% of full load; starting torque = 70.6/23 or 306% of full load. These ratios help to define the motor type.

the bars to the flux from the stator. To maintain the same (or adequate) rotor current, the motor must slow down so that more voltage can be generated to drive the current through the higher resistance cage. Hence increasing the rotor resistance reduces the full-load speed and causes increased I^2R losses in the rotor. If we can imagine a motor with high enough rotor resistance to give a 50% slip, then 50% of the rotor power input would be lost in heat. This is rarely desirable.

5-13 Means of Changing Rotor Resistance

The squirrel-cage rotors as described, once constructed, are of fixed resistance except for the effect of heat. In most cases it would be ideal to use a rotor which had a high resistance at the starting period for high torque and reduced current, and then a low resistance in the normal running region for reduced slip and losses. This can be done in two ways.

Wound-Rotor Motors. Instead of using the simple copper bars and end rings welded together, a two- or three-phase winding of ordinary insulated magnet wire is placed in the rotor slots. To avoid complexities, we can imagine this winding as being layed out exactly like the stator winding. Say it is Y-connected and the three ends are connected to slip rings on the shaft. Brushes that bear on these rings are connected to an external resistor (see Figure 5-20). By varying the resistance, we can obtain high starting torque as shown in Figure 5-21. Once the motor is up to speed, the resistance is cut out, so that the end rings are effectively short-circuited, and the motor operates at normal speed, say, point *a* on curve 1 of Figure 5-21.

Such motors are frequently used on hoists and cranes. They are relatively expensive because of the rotor winding, brushes and slip rings, and the need of an external resistor.

Variable speed can also be obtained with wound-rotor motors if the external resistance is sturdy enough for continuous operation. The exact speed at any load will vary with the nature of the load. Does the load take less torque at reduced speed? This is relatively common in industry. Then the torque required by the load might be as shown by line *b*.

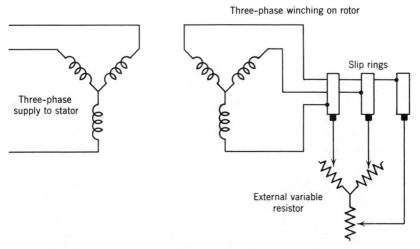

Figure 5-20. Schematic arrangement of windings and external resistance in a wound-rotor induction motor.

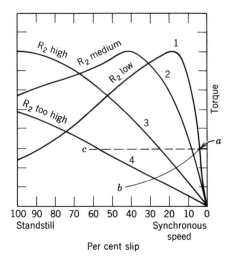

Figure 5-21. Speed-torque curves of a wound-rotor induction motor with various values of rotor resistance. It would be possible to build a squirrel-cage rotor that would duplicate any one of these speed-torque curves.

Now at setting 4 of the resistor, the motor operates at 30% slip; at setting 3 it operates at 17% slip, and so on.

Now imagine a load in which the torque required stays constant regardless of the speed. This is pictured as dotted line *c*. Now at resistor setting 4 the slip is 58%, at setting 3 it is 25%, and so on. The point is that, as the nature of the load changes or the magnitude of the required torque changes, the motor will vary in speed unless an operator (or some automatic control) corrects it. Hence this is a *varying-speed* type of motor.

One favorable point about these motors rests in the fact that although, say, 30% of the power transferred across the air gap is lost as heat, if the slip is 30%, most of that heat occurs in the external resistor and does not add to the temperature rise of the motor.

In this wound-rotor motor, as in any squirrel cage motor, changing the rotor resistance *does not change the maximum torque* until R_2 becomes higher than the value that makes the starting torque equal to the maximum.

5-14   Modern Rotor Construction

Up to this point we have considered the squirrel-cage rotor as being fabricated of copper bars and end rings, soldered or welded together.

For motors from fractional horsepower up to about 100 hp, this method of construction is practically obsolete. The inexpensive method of construction is to place the required stack of laminations in a mold which has cavities at the ends to provide for suitably sized end rings, sometimes with fans as an integral part of the mold. Molten aluminum is injected at high pressure into the mold; as a result, both end rings and all of the bars constitute a single casting integral with the rotor laminations.

Such a completed rotor with fan blades cast on one end ring can be seen in the cutaway view of the motor of Figure 5-22. No rotor bars are shown because they are buried in the laminations, probably about 0.010 or 0.015 in. under the surface (see Figure 5-23).

Multiple molds are made so that in one operation anywhere from two to eight rotors might be obtained.

The die-cast rotor is turned or ground to size (so as to leave the required air gap when placed in the stator) and then heated. The differ-

Figure 5-22. Cutaway view of an integral-horsepower, three-phase motor, showing aluminum die-cast rotor with integral fan blades cast on one end ring. Auxiliary fan and baffles are also used for ventilation. (Courtesy of Robbins and Myers, Inc.)

Figure 5-23. Squirrel-cage winding for induction motor. This element was originally cast into laminated steel core, which was then dissolved. Note the twist of the rotor bars. This is called "skew."

ence in expansion of steel and aluminum separates the latter from such intimate contact with the steel and minimizes stray currents. Since aluminum has only about 55% of the conductivity of copper, bar and end-ring sections of cast-aluminum rotors are larger than in equivalent copper constructions.

5-15   Another Method of Changing Rotor Resistance

It has been mentioned that an induction motor would be improved if it had a high-resistance rotor at the start which changed to low resistance when running. We saw one expensive way of doing this with the wound rotor and external resistance.

A much less expensive method involves what is known as a *double-cage rotor,* invented by Dolivo Dobrowolsky[2] (see Figure 5-24).

At this instant of starting such a motor, the rotor is exposed to stator frequency. The lower cage, being so deeply embedded, has high reactance and carries very little current. The rotor starts then with a resistance

---

[2] The reader is doubtless aware of the Soviet claims of having invented everything ahead of the Western world. Here is something that perhaps a Russian did invent. Unfortunately, the author can find no biographical data on Mr. Dobrowolsky. Perhaps he was a Pole.

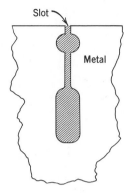

Figure 5-24. Cross section of a slot in a double-cage rotor. Before the advent of die-cast rotors, this construction with one bar in the top and another in the bottom was an expensive construction. Today the entire opening is filled with an aluminum integral with an end ring that contracts the depth of the rotor bar. Note that the top cage is of high resistance (low cross section), compared with the lower cage.

represented by the outer cage. As it gains speed, lower-cage current increases, until at running near synchronism the entire cross section of the bar is a low-resistance conductor, allowing very little slip.

Now refer back to Figure 5-17. This 50-hp motor was built with a double-cage rotor. Note that the full-load speed was 1175 rpm instead of the nominal 1140. We shall note later that this is what National Standards define as a "Design C" motor.

5-16 Effect of Voltage on Torque

If we reduce the voltage on a polyphase motor, the air-gap flux reduces in proportion. This, in turn, reduces the voltage induced in the rotor bars at a given speed. Hence less current flows in the bars with less flux built up around them to react with the field. The result is that the torque reduces as the *square of the voltage* because both factors that react to produce torque are affected.

Figure 5-25 shows the speed-torque curves of a motor supplied by a transformer with low-voltage tapes of 80, 70, and 60% of rated voltage. The entire range of the speed torque curve is affected. This is sometimes used for "easing" large motors onto a line without taking excessive starting current. It can be used for a small degree of speed control, although the supply transformer is comparatively costly.

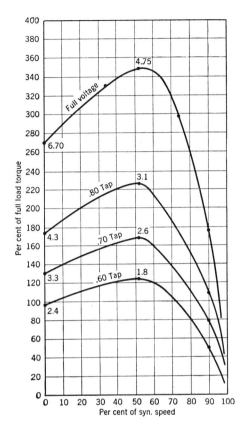

Figure 5-25. Speed-torque curves of a motor operating at rated voltage and at 80, 70, and 60% of rated. The numbers on the curves indicate the multiple of full load current under each condition. Thus the starting current at rated voltage drops from 6.7 times full load to 2.4 times full load if only 60% of rated voltage is applied.

5-17   Another Method of Speed Control

Because the polyphase induction motor operates basically from a revolving flux in the air gap, fixed by the frequency and number of poles, there seems to be little that can be done, inexpensively and efficiently, to change the speed of a given motor. We have seen that a wound rotor can be used or that a variable-voltage supply is effective within a small range. Aside from this, the only means of speed change is to tolerate wide ranges of speed (a) placing several sets of windings for different poles in the slots or (*b*) reconnecting special windings that are capable of being connected for, say, two or four poles.

To be able to place two sets of windings in the stator slots, naturally means that the wire size will be smaller and less capable of carrying heavy current without undue overheating. Hence two-speed motors of this sort are likely to be larger than single-speed motors of the same horsepower.

An example might be an oversized hoist motor wound for four and 24 poles. The four-pole speed is used for rapid raising or lowering the load, the 24-pole motor is switched on for slow and accurate positioning. Because they both handle the same load, the torque of both motors must be equal, but, as the speed of the 24-pole motor is much less, its *rate* of doing work is reduced as is the horsepower for which it is designed.

In general, three lines of multispeed motors are available, variable torque (very low torque at low speed, suitable for a fan or blower), constant torque at either speed, or constant horsepower at either speed. The latter designs are somewhat oversized, because the slower the motor, the larger its frame size. This accounts in part for the popularity of higher-speed motors with built-in gear reducers when low-speed drives are required.

5-18   Applications

In the remainder of this chapter, we shall deal with problems connected with the applications of polyphase induction motors, starting naturally with the simplest case.

5-19   Horsepower and Speed Required: Simple-Duty Cycle

If you have made some basic changes on a machine tool, or developed some new gadget that you wish to drive by an electric motor, it is conceivable that the horsepower required might be completely unknown. There is no need to be mysteriously scientific about this solution. Borrow a motor, an ammeter, and a voltmeter by some means or other and drive the device. Note the amperes required at normal voltage when the device is working. Suppose the current is 20 amp per line and the motor name plate reads 15 amp at 220 v. This latter value is about the usual full-load current for a 5-hp motor at that voltage. If you suspect that the motor will operate intermittently or that there is no possibility that the device will be overloaded, this 5-hp motor might be satisfactory.

If, on the other hand, you wish to be conservative, a $7\frac{1}{2}$-hp motor with a usual full-load current of about 22 amp should be selected. The correct motors for thousands of machines with simple "duty cycles" are selected in just such a manner. It is assumed here that these are industrial applications with motor "out in the open" and unrestricted ventilation can be obtained.

5-20   Machine Drive with Frequent Starting

Instead of the foregoing application in which the load is fairly constant and the machine is likely to be operated for fairly long periods, consider the following case:

A motor is required to start, come up to speed, operate at 4.2-hp output at about 1725 rpm for 1 min, coast to a stop, and then start again, repeating this cycle 30 times/hr.

Here is a simple case of a duty-cycle application that can take on a variety of combinations, many of which are complex enough so that an engineer from the motor supplier is needed as a consultant. In many cases the motor may have to be especially designed for your machine operation. (In this latter case it is hoped that you will have a large enough potential market to justify all of the work involved.)

To the novice, the foregoing description looks like a simple case of applying a four-pole 5-hp motor with the expectation that it will operate satisfactorily without overheating. This might be the case, but the doubt is caused by the frequent starting of the motor, during which an unusual amount of heat is developed.

The first thing that the motor salesman wants to know, aside from the horsepower, is the inertia of the driven parts. This is the "flywheel effect." It can be measured as WR^2, where W is the weight of the rotating parts and R is the radius at which the weight is apparently operative. This radius is called the *radius of gyration* (see Figure 5-26).

The next step is then to supply a motor for the project and measure very accurately the time required to accelerate the load to final operating speed. If this trial motor is then returned to the motor engineering laboratory, flywheels of known WR^2 can be tried until a combination is found which results in the same accelerating time as that observed with the customer's machine.

To consider numerical values; the WR^2 of the motor rotor, shaft, and fan is known to be 4.0 lb-ft^2. Flywheels were added to the motor

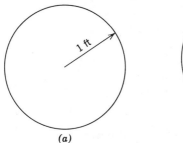
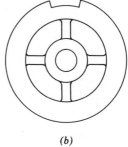

(a)   (b)

Figure 5-26. If (a) represents a uniform cylinder, the effective radius at which the weight apparently operates is 70% of the two radii. This is the "radius of gyration," here 0.7 ft. If the cylinder weighs 10 lb, the inertia as WR^2 is 4.9 lb-ft.² The inertia of the wheel shown in (b) can be calculated but it is a very tedious process.

shaft until it was found to accelerate in 1.8 sec as it had on the customer's machine. The added WR^2 was 5.1 lb-ft². This, then, represented the value of the machine inertia.

Suppose, however, that the machine load was belted to the motor so that the machine operated at 670 rpm. In any case the inertia of the load or parts of the load must be referred to the motor and motor speed.

Hence the actual inertia of the machine would be

$$\left(\frac{1725}{670}\right)^2 \times 5.1 \text{ or } 34.0 \text{ lb-ft}^2.$$

Because this inertia is operating at a lower speed, it affects the motor as though it was the lesser value of 5.1 lb-ft². The motor "sees" only this value. Note that the correction will always vary as the square of the speed ratio.

A second method of determining the time of acceleration or the WR^2 of the total system involves knowing the exact shape of the speed-torque curves of the motor and of the load. This is done by using the following formula step by step for, say, every 100 or 200 rpm (see Figure 5-27).

$$\text{time} = 0.00325 \, \frac{\text{change in rpm} \times WR^2}{T \text{ available for acceleration}}. \tag{5-10}$$

If we assume a WR^2 of 1, go through this process, and find that the time taken to get up to speed is 0.5 sec, but the test shows 3 sec to accelerate, the WR^2 of the load must then be 6 lb-ft².

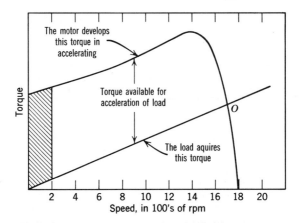

Figure 5-27. The difference between the torque developed by the motor and the torque required by the load is available for acceleration. By determining the average height of the cross-hatched area it is possible to calculate the time required to accelerate from 0 to 200 rpm. This can be done up to the final operating point *O*, adding step by step.

5-21 Heating During Starting

Why all of this emphasis on WR^2? It is because it is vital in determining how hot a motor will get when started and reversed.

The heat generated in the rotor during starting is exactly equal to the energy stored in the rotating parts. The latter is a function of WR^2.

This cannot be emphasized too strongly. It makes no difference whether you design a motor with high starting torque which starts a load in 0.1 sec, or design with such lower starting torque that it takes 1.0 sec to accelerate the same load. In either case, the energy lost as heat in the motor is *exactly the same*.

This is not an isolated phenomenon in nature. If you are driving your car at a given speed and apply your brakes so that the car comes to a stop in 10 sec, a certain amount of heat is developed in your braking system. Stop it in 5 sec and exactly the same amount of heat is generated. The heat (energy) depends upon the initial speed and weight of your car, which, in turn, measures the energy stored in it by its motion. Here we are dealing with deceleration instead of acceleration, but the same principles apply.

Energy can be expressed in many units. Heat as energy might be expressed in Btu per hour, as in your furnace, or in electrical terms as kilowatt-hours, which is what you pay for on your electric meter.

But it is convenient here to express the energy stored in a rotating motor and its load in terms of watt-seconds (wsec), since this will relate directly to motor losses.

So, without derivation, we point out that during acceleration,

$$\text{rotor wsec} = 0.00744 \times \frac{WR^2}{g} \text{ (synchronous rpm)}^2 \times (s_1{}^2 - s_2{}^2), \quad (5\text{-}11)$$

where g = acceleration of gravity or 32.2, which appears here because of the units used,

s_1 = initial slip, which at starting is 100% or 1,

s_2 = final slip (usually assumed to be 0 for simplicity in simple cases of starting).

5-22   Things the Motor Supplier Must Know

We have seen from the foregoing that the motor user should supply information on:

The desired motor speed.

The horsepower of the load.

The speed and WR^2 of each moving part.

The time that the motor will be loaded and the number of starts per hour.

In addition, to supply a motor the motor manufacturer must know about his product:

The WR^2 of his rotor, shaft, and fan.

The relative resistances of the stator and rotor.

Watt loss at various loads at which the motor might operate.

Temperature rise of the motor at the foregoing loads.

5-23   Continuation of Example

This motor operates at 4.2-hp load for 1 min, coasts to a stop, and then starts again, going through this 30 times per hour, displays the following data:

$$R_1 \text{ per phase} = 0.658 \ \Omega,$$
$$R_2 \text{ per phase} = 0.51 \ \Omega.$$

From a brake test:

Horsepower Output	Watt Loss
5.5	740
5.0	655 (temperature rise at this load 42°C)
4.2	530
4.0	520
3.0	430
0	240

Total WR^2 in motor-speed terms is 9.1 lb-ft².

Energy appearing as heat in the rotor during starting:

$$\text{rotor, wsec} = 0.00744 \times \frac{9.1}{32.2} \times 1800^2 \times 1 \text{ (from Eq. 5-11)}$$

$$= 6850 \text{ wsec,}$$

$$\text{ratio of } R_1/R_2 = \frac{0.658}{0.51} \text{ or } 1.29,$$

stator, wsec = 1.29 × 6850 or 8830 (all three phases).

Total heat generated in the stator and rotor during starting is 6850 + 8830 or 15,680 wsec. This ignores the iron loss present during this time.

Now refer to Figure 5-28 to obtain a graphic picture of the operation.

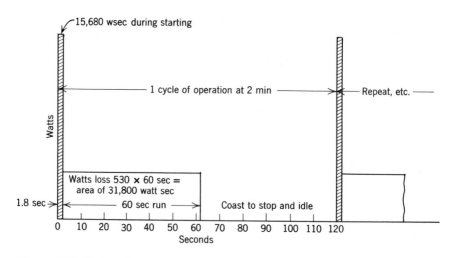

Figure 5-28. During the 1.8 sec of starting, heat energy of 15,800 w sec is dissipated in the motor, as compared to 31,800 w sec during a full minute of operation at 4.2 hp.

We know from the test data that the losses at 4.2-hp output are 530 w and, as it runs at that load for 60 sec, the heat energy lost during the load run would be 530×60 or 31,800 wsec.

Will the motor operating hour after hour on this duty cycle burn up or not? To check this we must have some point of reference for operating temperature. We have it by knowing that, at 5 hp, with a loss of 655 w, the motor rose in temperature only 42°C.

$$655 \text{ w, } 120 \text{ sec} = 78{,}600 \text{ wsec.}$$

Hence, in any 2 min of normal operation 78,600 heat units dissipated in the motor will raise its temperature 42°C. During this cycle of operation, calculated in the foregoing, the total watt-seconds dissipated are

$$\begin{array}{r} 15{,}680 \\ 31{,}800 \\ \hline 47{,}480 \text{ in 2 min.} \end{array}$$

The temperature rise will vary proportionately to the heat generated and, hence, it would be

$$\frac{47{,}480}{78{,}600} \times 42°\text{C or } 25.4°\text{C.}$$

(More detailed explanation of losses, temperature rise, and ventilation will be given in Chapter 14.)

We can conclude that the motor would be suitable for this application. It might even be possible that a 3-hp rating would be satisfactory if it had enough maximum torque to handle a 4.2-hp load.

Suppose, however, that the watt-second loss in the two-min cycle had been, say, 100,000. Now the motor would be required to operate at a temperature of

$$\frac{100{,}000}{78{,}600} \times 42°\text{C or } 53.5°\text{C.}$$

This could be excessive. What can we do if the results show excessive losses on a given duty cycle? Put on a larger motor, say, $7\frac{1}{2}$ hp? This could be worse, because a $7\frac{1}{2}$-hp motor is sure to have a higher WR^2 than a 5-hp design, perhaps twice as much.

Possible solutions include lightening the rotor by using laminations with a series of holes punched out, redesigning to reduce the resistance of the stator winding, or redesigning with a set of laminations in which the rotor diameter is less.

5-24 Methods of Rapid Stopping

In the foregoing example it was assumed that the motor would coast to a stop in about 1 min. Suppose it is necessary in the machine operation that the motor must come to a complete stop, and in automatic or semi-automatic machines perhaps stop as quickly as possible to speed up operations. This can be done in one of three ways.

1. With a magnetic brake, as described in Chapter 15.

2. By disconnecting the ac leads and applying a relatively low-voltage, high current dc supply to two of the lines. This results in a generating action that causes braking to a stop.

3. By "plugging." We know that, if two lines of a three-phase motor are reversed, the motor reverses. Suppose a motor is operating at full speed forward. If two of the leads are reversed (usually by a magnetic contactor), the air-gap flux immediately starts rotating in the direction opposite to the rotor. The resulting generator action brings the rotor to a rapid stop and then quickly accelerates it in the opposite direction. This is called *plugging*. But if we do not want it to *reverse*, then a special control can be used that opens the circuit as soon as the motor is stopped. This is a relatively inexpensive method of quick stop, but, as in the case of adding the direct current to the leads, it generates considerable heat in the motor.

Again WR^2 is important in determining the watt-seconds of heat. In fact, (5-11) still holds if we take account of the fact that the slip now goes from 2 (backward synchronous) to 1 (standstill). Then

$$s_1{}^2 - s_2{}^2 = 3,$$

and (5-11) becomes

$$\text{rotor, wsec} = 0.00744 \times \frac{WR^2}{g} \text{(synchronous rpm)}^2 \times 3.$$

5-25 Example of Plugging to a Stop

Consider the same duty cycle and motor of Section 5-23.

The rotor watt-seconds during acceleration were 6850. To plug to a stop, we can see by comparing equations that the value will be three times as great or

$$\text{rotor, wsec} = 3 \times 6850 = 20{,}550 \text{ wsec,}$$
$$\text{stator, wsec} = 1.29 \times 20{,}550 \text{ or } 26{,}490 \text{ wsec.}$$

The total watt-seconds for stator and rotor owing to plugging to a stop are 26,490 + 20,550 or 47,040 wsec.

Then, from Section 5-23, in watt-seconds,

accelerating	15,680
running 1 min	31,800
plug to stop	47,040
total	94,520 wsec.

Watt-seconds that produce a 42°C rise in the same period are 78,600.

$$\text{probable temperature rise of motor } \frac{94{,}520}{78{,}600} \times 42°C \text{ or } 50°C.$$

If the motor is insulated to be able to operate satisfactorily at 50°C, the application is suitable; if not, one of the redesign methods previously described must be tried.

It will be noted from the foregoing that it is unnecessary to go through any further calculations once we have calculated the watt-seconds for acceleration. Multiply by three to obtain the watt-second for plugging to a stop and multiply by four (through the changes in $s_1{}^2 + s_2{}^2$) to obtain the watt-seconds when plugging to reverse the motor.

These comparatively large values of motor loss and the attendant heating often make external electric brakes useful, as compared to plugging.

5-25   The Application of High-Slip Motors

Most integral horsepower induction motors used on simple applications under normal atmospheric conditions with 5% slip or less are called *general-purpose motors*. But the industry also manufactures a line of high-slip motors, generally available from stock. One line has slips of from 5 to 8%, another from 8 to 13%. Considering that higher slip adds to rotor losses and to temperature rise, we might wonder as to their usefulness. They are applied chiefly to reciprocating loads, such as deep-well oil pumping, punch presses, and the like. Their usefulness depends upon the load having high inertia, either by the nature of the equipment or through an added flywheel.

Consider the current or horsepower requirements of a periodically varying load, as shown in Figure 5-29. These large pulsations of current may cause disturbances in the supply lines, affecting other equipment. To consider the effect of a high-slip motor, we must give consideration to the energy stored in a rotating (or in general a moving) mass. Suppose a flywheel of 1000 lb is rotating at 600 rpm. It represents a certain

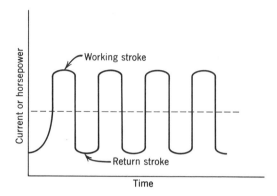

Figure 5-29. Representation of a reciprocating load showing variations in horsepower or current input. Current, as shown on a recording ammeter, rarely results in such clean regular shapes as indicated here.

level of energy. To raise it to 800 rpm requires a definite increase in energy supplied. Lower its speed to 500 rpm and the lower level of energy represented means that some energy must be given out, either into a brake that slows it down or into the attached mechanical load.

This effect of feeding energy into a moving system when it speeds up and having energy fed back into the system when it slows down depends naturally upon the change in speed; hence the value of a high-slip motor.

A typical oil-well pumping application is shown in Figure 5-30. Its current input as the pump beam works up and down might resemble that shown in Figures 5-29 or 5-31, but note in Figure 5-31 that the current curve is smoothed out by the use of a high-slip motor. When the heavy load stroke comes on, the motor slows down enough so that the energy stored in the heavy moving parts helps to do some of the work. Similarly, on the light load portion of the operation, the motor speeds up, so that it supplies not only the actual work done but the additional power required to restore the mass to higher speed. Hence the motor's light-load period becomes not so light a load and its "valley" is lessened.

As the I^2R losses vary obviously as the square of the current, these losses at the peak loads rise more above the average than they reduce during the light load. Hence a smoother current input curve can reduce the losses.

Oil producers think in terms of kilowatt-hours per barrel pumped. Extensive tests show that this cost is reduced when high-slip motors are used.

Figure 5-30. A typical oil-well pumping unit with a high slip motor drive. As the beam moves up and down, it moves with it several thousand feet of rod which actuate the reciprocating pump at the bottom of the well. Extra weights are used to balance the weight of the pump rods. Courtesy of *The Oil and Gas Journal,* Tulsa, Oklahoma.

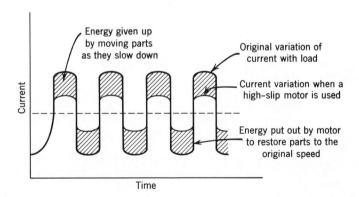

Figure 5-31. Curves showing how a high slip motor with the speed varying with the load smooths out the load pulsation. This effect also varies with the mass of the moving parts. On some punch presses, for instance, fly wheels are added to aid in the leveling of the load.

5-26 A Flash of Genius

This has been a long chapter, but, after all, 90 to 95% of the motors used in industry are of the polyphase induction type. The machine deserves a thorough exposition.

Dozens of books have been written on this type of motor as well as hundreds of technical papers. They deal with methods of calculating certain factors that affect the performance or quantities that are necessary for the accurate design of motors to yield predetermined characteristics.

We might imagine that the invention of the polyphase induction motor was the result of months of experimental work and/or mathematical gymnastics.

Not at all.

Nikola Tesla (1856–1943), educated in Hungary, came to this country in 1884. While taking a vacation at the seashore, he sketched out the stator, the placement of the windings, and other diagrams in the sand. He satisfied himself that a rotating field would result in which a squirrel cage would produce torque. His patent was granted in 1888. The hard analytical work came later by a number of experimenters.

Patent rights were sold to George Westinghouse. Tesla invented a few other devices of minor importance. For the rest of his life he experimented with ideas, such as transmitting power without wires, that never materialized. On his 78th birthday he announced to the press that he had invented a death ray that would immobilize 10,000 airplanes 250 miles away or kill 1,000,000 men instantly. His later years are best forgotten. He might be called a "single-shot" genius.[3]

---

[3] Tesla also invented the "Tesla coil." This consisted of a conventional high-voltage transformer, say about 20,000-v output, which was connected to a capacitor, a spark gap, and a few turns of wire. Inside these turns of wire was another coil of hundreds of turns. The latter formed an air-cored transformer. In operation this latter coil was the secondary of a transformer which put out, say, 500,000 to 1,000,000 v. The circuit went into oscillating resonance at about 500,000 cycles.

The only commerical use for this device (in the knowledge of the author) was in carnivals, where one could see "the death-defying little lady pass 1,000,000 v through her body."

By touching one terminal of the secondary with one hand, one could obtain a "brush discharge" of many sparks from the other hand or a long crackling spark to any grounded object.

High-frequency currents tend to travel close to the surface of a conductor. In the case of these several hundred-thousand cycles the current never penetrated the skin deeply enough to reach the nerve endings. The little lady felt nothing at all except a slight prickling sensation as the spark jumped.

Chapter 6

The Plain Single-Phase Motor

In dealing with polyphase motors, we had the advantage of sources of voltages which reached their peaks at different times. These were 90° apart for two-phase motors, and 120° apart for three-phase motors. Applied to windings suitably spaced about the stator of a motor, this combination of *time* and *space* displacements resulted in a rotating magnetic field in the air gap.

A single-phase supply, of course, consists of only one alternating voltage between the two supply lines.

6-1 Single-Phase Winding and Flux System

A stack of laminations identical to those used for a polyphase motor is shown in Figure 6-1. Two sets of two coils are shown in the slots, making this a two-pole winding. The coils of any one pole are concentric, the inner coil having a shorter pitch than the outer coil. Although only two coils per pole are pictured here, three, four, or five could be used, depending on the size of the motor and the number of slots. This, then, is a concentric type of winding. The coils of any one pole are connected in series. These can be in series of parallel with those of the other pole, depending upon the design.

Connect these coils to an alternating voltage, and an alternating flux will be set up through the laminations and across the air gap approximately as shown. It pulsates in one axis only (in this case vertically). It goes from zero to a maximum flux, upward, back through zero to a maximum flux, downward, and then to zero again in one cycle. There is apparently no rotation of flux in this system at all.

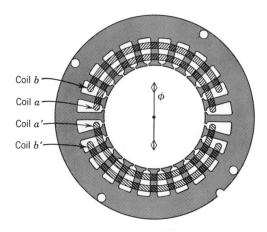

Coil b
Coil a
Coil a'
Coil b'

ϕ

Figure 6-1. Main windings of a two-pole, single-phase motor. Coils a and b are connected in series. Coils a-b and a'-b' may be in series or parallel. These are concentric windings.

If we place a squirrel-cage rotor in the center of this stator, it will buzz and get hot but will not turn.

However, if we should give the shaft a slight spin in either direction, it will immediately pick up in speed and run. We are faced with the unfortunate fact that a single-phase motor of the induction type has no starting torque. A number of methods are used to start these motors, and the motor then takes its name from the starting method used. Thus we have split-phase motors, capacitor-start motors, etc. These will be discussed in detail in the next chapter.

With only a pulsating flux available, how can we account for the fact that it runs at all?

There are two theories, each with its own body of mathematics for analyzing this action and calculating the performance. They are complex, unwieldy, and generally deplored. Neither is easy to explain in words.

6-2   The Doubly Revolving Field Theory

Consider two oppositely rotating vectors, each of equal length and each operating at the same speed. Consider them at the time a in Figure 6-2. The sum is zero.

An instant later they have rotated to the positions b-b. The sum is shown as B.

When the two vectors coincide in a vertical position, their sum is

C which is twice the magnitude of either component. As they continue rotation past the vertical, they have crossed over each other and reduce again to B and then zero.

Continued rotation brings them below the horizontal and to resultant values of B^1 and C^1. Thus we can see that two oppositely *rotating* vectors can result in a *pulsating* vector in one plane, varying in magnitude from zero to twice that of the components.

This is the basis for one theory of single-phase motor operation. It assumes that the pulsating flux in one plane is really the resultant of two oppositely rotating flux systems. At standstill the fields are equal and opposite. The motor has no tendency to turn.

Once it is made to rotate, current induced in the rotor bars reacts strongly with the flux turning in that same direction. The two react to produce increasing accelerating torque. On the other hand, the current and flux system set up in the rotor works against the oppositely rotating

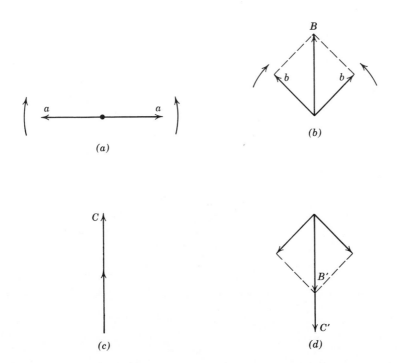

Figure 6-2. Two oppositely rotating vectors can result in a pulsating vector in one plane. (*a*) Two oppositely rotating vectors at this instant add up to a vector sum of zero. (*b*) An instant later when they have reached points *b-b*, the sum is *B*. (*c*) When the two vectors coincide their sum is *C*.

flux system and weakens it. Thus the oppositely rotating flux becomes weaker and weaker as the motor picks up in speed. It never disappears, but even at maximum speed its presence is noted as a drag on the rotor which causes extra losses and shows up as a 120-cycle torque pulsation.[1]

6-3   The Cross-Field Theory

A second approach is used to explain why a single-phase motor runs once it is started. It too is rigorous in its mathematics. It is called the *cross-field theory*.

Consider the same motor, with its elementary two-pole winding as shown in Figure 6-1. For our present purposes we wish to consider in more detail the current set up in the rotor bars, and so we shall reduce the physical structure to the more simple schematic diagram of Figure 6-3.

[1] The novice, first exposed to the concept that the pulsating flux consists really of two oppositely rotating fluxes, usually looks upon this as a graphical and mathematical trick which has no actual reality. But this concept is mathematically rigorous, and explains the facts that the motor is willing to operate in either direction as well as the 120-cycle pulsation, which is the natural outcome of a flux rotating oppositely from the rotor.

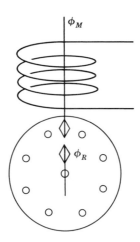

Figure 6-3. Schematic diagram showing the stator flux ϕ_M, the voltages induced in the rotor bars (when stationary) and the short-circuited rotor currents which build up an opposing flux ϕ_R. This is ordinary transformer action.

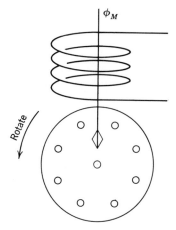

Figure 6-4. As the rotor turns through the field, generator action causes voltages in the bars as shown.

At a given instant the flux of the stator field is represented by ϕ_M. Because it cuts the stationary rotor bars, it induces voltages in the directions as shown. The end rings complete the circuit so that currents flow from left to right in this figure, setting up a counterflux ϕ_R. This is ordinary transformer action as previously described, with the secondary (rotor) apparently short-circuited. If we examine these current directions with reference to the air-gap flux from the standpoint of *motor action,* it will be seen that the bar currents on one side of the rotor give torque in one direction, those on the other side oppose it.

Now consider that the rotor is turning. While the air-gap flux is pulsating, think of it at one instant as it is shown in Figure 6-4 and consider dc *generating action.* At the instant shown these rotor bars, rotating counterclockwise, generate emfs coming out of the top of the rotor and entering the bottom. The end rings complete the circuit and the rotor is an apparent "winding," as indicated in Figure 6-5. These rotor currents build up a flux system as shown by ϕ_C.

We now have an important concept that can explain why the motor rotates. Think back to the two-phase motor. Two stator windings displaced 90° were excited by two voltages also 90° out of phase in time. By this theory, in the single-phase motor, the one stator winding gives us a flux system which is vertical. Generator action in the rotor, once it is turning, builds up a flux system in the horizontal axis of the rotor. This flux path includes not only the rotor but also the air gaps and the stator teeth and yoke. We have, then, two flux systems 90° apart in space. The inductance of a rotor circuit is such that the two currents are about 90° apart in time.

The cross-field theory explains the single-phase motor as an approximate two-phase case in which the second-phase flux originates in the rotor.

6-4   Direction of Rotation, Losses, Vibration

Note in Figure 6-5 that it is assumed that the motor is rotating counterclockwise. As a result, generator action builds up the cross flux to the right as shown. But had the motor started in the opposite direction, the generated voltages in the bars would have been reversed and the cross field would be toward the left. This also explains why the motor is willing to operate in either direction. Had this really been a two-phase motor and we had reversed the connection of the second phase, its flux would have reversed exactly as previously described and the motor would have changed rotation.

Generating these emfs and currents to set up the cross field requires power, which comes from the torque of the rotor. Hence they become a drag on the rotor equal to that caused by the backward flux when considering the doubly revolving field theory. Theoretically, a single-

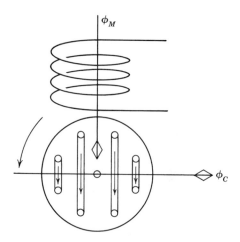

Figure 6-5. Because of the symmetry of the rotor bars and end rings, we can consider that each bar above the horizontal axis is connected to one below. Thus, in a sense, the rotor is a "coil winding" with its magnetic axis horizontal. It builds up a cross flux ϕ_C.

phase motor is never "unloaded." Even though there is no mechanical load on its shaft, it is "loaded" by having to generate its cross field or by the retarding torque of the backward flux. Hence it never comes as close to synchronous speed as does a poly phase motor.

The cross-field flux is pulsating. The rotor bars also rotate through it. The result is a frequency of 120 cycles in the rotor, causing pulsations in the torque.

Regardless of which theory is used, we end up with the unfortunate fact that the torque of a single-phase motor varies from position to position as it turns. Mechanical parts of the motor that can respond to this 120-cycle pulsation (or multiples of it) have the unpleasant habit of doing so. Special resilient mountings help to keep this vibration from spreading throughout the driven system. Some types of single-phase motors are worse offenders than others.

Chapter 7

Types of Single-Phase Motors
Identified by Starting Methods

We have seen in Chapter 6 that the single-phase motor with only one set of windings is capable of running once it is started. The various means of obtaining this starting torque will now be described.

7-1 Split Phase

Figure 7-1 shows a typical stator for two-pole operation in which two sets of windings have been inserted. The winding to be used when the motor is running is called the main winding. With its squirrel-cage rotor in place, assume this winding is connected to a 115-v 60-cycle source. The current and voltage waves will take relative positions, as shown in Figure 7-2, more conveniently represented by the two phasors.

We expect that the current would lag behind the voltage because the coils embedded in the steel naturally build up a strong magnetic flux which gives the coil considerable inductive reactance.

Now examine the second winding at right angles to the main. This we call the *start* or *auxiliary* winding. If this winding were identical with the main and connected to the same single-phase source, it would build up a flux displaced by 90° from that of the main winding. The two flux systems would combine into one, with its axis midway between the two sets of windings. No further rotation would occur. The motor would hum and get hot.

But instead we provide the auxiliary winding with finer wire having a high resistance. Fewer turns are used. This reduces its inductive reactance. Thus the auxiliary winding has high resistance and low reactance, and its current does not lag so far behind the voltage.

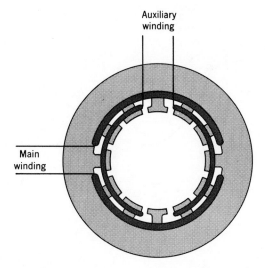

Figure 7-1. A two-pole, single-phase stator with auxiliary and main windings 90° apart. If arranged for four-pole operation, the main and auxiliary windings would be 90 electrical degrees but only 45 mechanical degrees apart.

The diagram of Figure 7-3 shows the voltage (the same for each winding) and the currents in the main and auxiliary windings. Note that the two currents are out of time phase, not by 90°, as needed in a two-phase system, but by about 30°. This then forms a sort of imitation two-phase system which produces a rotating flux. We have

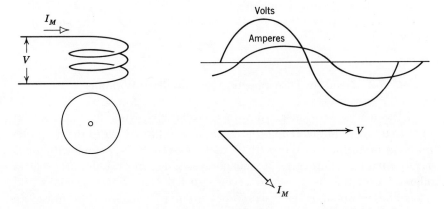

Figure 7-2. With the main winding connected to a suitable supply, the current, with the rotor stationary, will take up the position shown. Note that the main winding of Figure 7-1 is shown schematically.

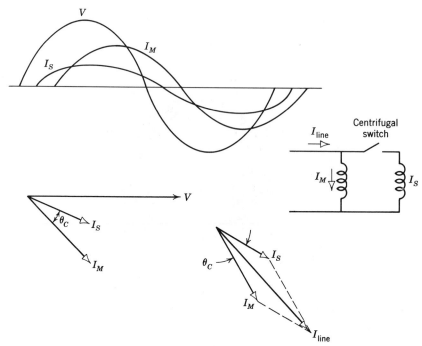

Figure 7-3. With both main and auxiliary windings connected to the line, the individual currents are as shown. The combined currents add to become the I line, which is the total locked rotor or starting current taken by the motor.

now "split the phase" of a single-phase source. Hence the name, more properly called *resistance split-phase.*

The total current taken by the motor at starting is the sum of I_s and I_m.

7-2   Need for Disconnecting the Auxiliary Winding

If the motor were operated with the auxiliary winding remaining in the circuit, it would be found that it did not add anything to the desirable running performance. Worse than that, because it is wound with relatively small wire to give it enough resistance, it would burn out in about 5 sec; that is, it would become so hot that its insulation would melt and it would either "short" or "ground." We are thus forced to consider some means by which the auxiliary winding, once it has started the motor, is disconnected from the line, but which connects it again when the motor is through with its operation and comes to a stop.

7-3   Centrifugal Switches

The centrifugal switch is a device consisting of an actuator and a contact plate. The actuator consists of hinged weights which can "fly out" a short distance when acted upon by centrifugal force. A spring tends to restrain this action. The speed at which these weights move out is fixed by the weights and the strength of the spring or springs. As they move out, a linkage with a sleeve on the shaft causes the sleeve to move axially. This sleeve, with the motor stationary, normally presses one hinged contact arm against another stationary contact. These two contacts are connected in series with the auxiliary winding. The motor starts and at about two-thirds of final speed the switch opens. The auxiliary winding is disconnected and the motor runs on its mail winding only (see Figure 7-4).

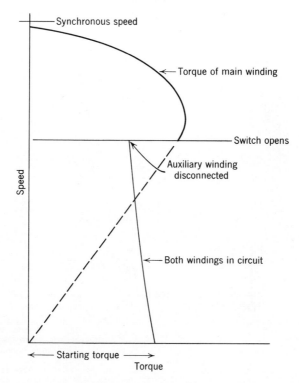

Figure 7-4. Speed-torque curve for a split-phase motor showing torque of combined windings and main winding only. In this case the "pull-up torque" (lowest value measured during acceleration) would occur when the switch opened.

Figure 7-5. A typical rotor with aluminum die-cast cage and centrifugal mechanism. Centrifugal force throws out the weights against the action of the springs. This also retracts the sleeve so that it moves to the right. (Courtesy of Robbins & Myers, Inc.)

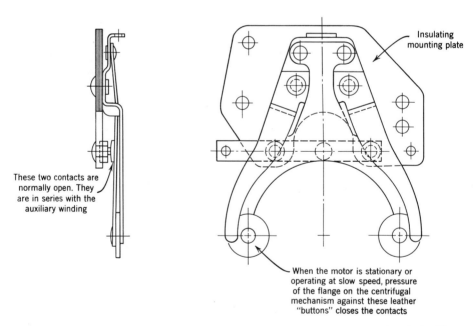

Insulating mounting plate

These two contacts are normally open. They are in series with the auxiliary winding

When the motor is stationary or operating at slow speed, pressure of the flange on the centrifugal mechanism against these leather "buttons" closes the contacts

Figure 7-6. A typical contact mechanism for use with a centrifugal switch. The friction between the leather "buttons" and the rotating sleeve lasts only for the fraction of a second while the motor starts.

120

A heavy overload can slow down the motor, causing the contacts to close again. The auxiliary winding, then being connected, adds its torque and the motor speeds up. This opens the contacts again. This cycling can repeat if the overload continues, causing burned contacts and/or a burned out auxiliary winding (see Figures 7-5, 7-6, and 7-7).

7-4   Current Relays

In addition to the centrifugal switch for removing the auxiliary winding from the circuit, a current relay is frequently used. This consists essentially of an electromagnet which actuates a contact arm. Its action is based on the fact that, when a motor is connected to the line, it draws a comparatively large current. The current gradually reduces as

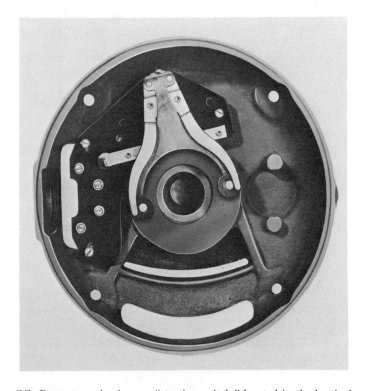

Figure 7-7. Contact mechanism, or "starting switch," located in the head of a motor. The screws at the left in the insulating mounting plate are the back of binding posts used as terminal connections for the motor windings. (Courtesy of Robbins & Myers, Inc.)

the motor gains speed. Thus it is the decay of current that permits the relay to open, disconnecting the auxiliary winding.

The relay coil is connected in series with the main winding. The contacts are in series with the auxiliary (see Figure 7-8). These contacts are normally open. Connecting the motor to the line closes the contacts by the pull of the electromagnet. As the motor gains speed, the current drops off and the contacts open.

These relays must be selected very accurately by the motor designer; otherwise the "dropout point" may be such as to disconnect the auxiliary winding too soon or too late.

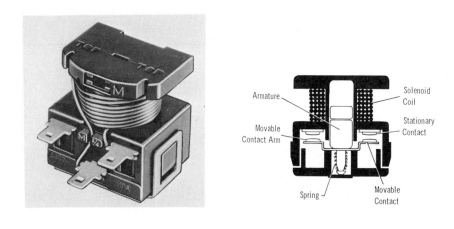

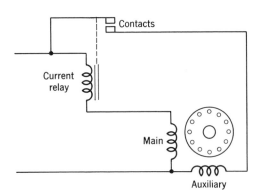

Figure 7-8. A current relay for motor starting. With the motor connected to the line, the heavy main winding current closes the contacts connecting the auxiliary winding to the supply. With pick-up in speed, this current reduces, opening the contacts. Current relay without protective cover. (Courtesy of Metals and Controls Division of Texas Instruments)

Refrigerator motors are usually built so that the active parts of the motor are exposed to the gas in the compressor. These are called *hermetic motors*. It has never seemed feasible to expose a centrifugal device and contactors to the refrigerant; hence these employ relay starting. They can be connected externally in the line.

7-5   Capacitor-Start Motors

In considering the split-phase motor, we found that torquewise we were handicapped by not being able to split the phases through a large enough angle. Ideally, if the currents in the main and auxiliary windings were equal and 90° apart, we could obtain all of the starting advantages of a two-phase motor.

This condition can be approximated by the use of a capacitor which,

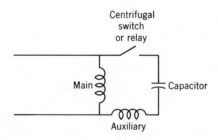

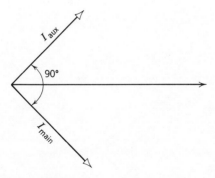

Figure 7-9. A capacitor added to the auxiliary winding circuit can be of such value that it will make the current of that winding equal to that of the main and lead it by 90°. The two windings then apparently have a two-phase supply at starting.

as we saw in Chapter 3, has the peculiar property of making the current lead the voltage. This becomes an extremely useful component to add to the auxiliary winding circuit (see Figure 7-9).

In practice it is not always necessary nor economical to bring about the true two-phase as shown. Figure 7-10 illustrates a condition that results in sufficient starting torque with fewer microfarads of capacity. In spite of this compromise, motors of $\frac{1}{4}$ to 10 hp might require from, say, 60 to 1000 μf.

7-6   The Electrolytic Capacitor

The fact that a single-phase motor could be started in this manner was pointed out by Steinmetz many years ago. But at that time only the foil-paper-foil construction was known for making a capacitor. These capacitors were bulky and expensive. A 1000-μf capacitor of this type would be considerably larger than a bread box.

About 1930 the etched-foil electrolytic capacitor was developed. By this construction it was possible to build, say, 60 μf in a case about the size of a flashlight battery, and not too expensively at that.

But it has one disadvantage. It cannot be connected permanently to the line. In fact, standards limit it to 20 starts per hour, each of 3 sec duration. There is a considerable safety factor in this, but, nevertheless, the capacitor-start motor must get its load up to speed quickly and then have its auxiliary winding and capacitor disconnected from the line. Either the centrifugal switch or current relay will do this.

Normally this does not cause much difficulty in application. The start-

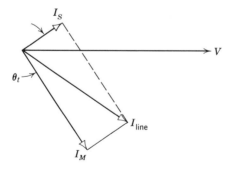

Figure 7-10. The capacitor start motor showing currents with the rotor stationary. Here fewer microfarads are used, but useful values of starting torque are still obtained. Note that because of the favorable positions of I_M and I_S, the line current is not excessive.

ing torque of a capacitor-start motor can readily be as high as 400% of the rated full-load torque. Furthermore, as this type of motor gains speed, it develops still more torque for accelerating the load. Hence, unless the motor load has unusual inertia, the starting and accelerating torques of this motor snap it up to speed rather quickly. This is a great improvement over the resistance split-phase motor with starting torques usually of about 100% of full load and the tendency for the accelerating torque to reduce as the motor picks up speed (see Figure 7-11).

The development of the electrolytic capacitor made all of this possible and greatly changed the complexion of the motor industry in the past 30 years.

7-7   The Permanent-Split-Capacitor Motor

We have seen that many microfarads of capacity were necessary to permit enough current to flow in the auxiliary winding to yield high starting torque.

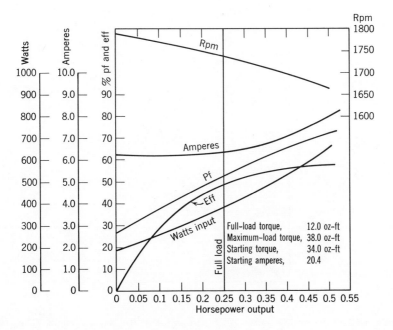

Figure 7-11. Typical performance of a ¼-hp 1725 rpm capacitor start induction-run motor. A split-phase motor of the same rating might have identical performance except for starting amperes and torque.

What would happen if we used only 4 or 5 μf of oil capacitors which could remain constantly in the auxiliary circuit? At starting we would have very little help from the small current that could flow through the winding and capacitor (see Figure 7-12).

The accelerating torque would be low, but the maximum torque would increase over that developed by the main winding only.

Suppose more capacitors are added to the circuit. This increases the benefits, up to a point, and then it can be shown that, although starting and accelerating torques continue to increase, the running performance becomes very poor. Noise, reduced speed, and overheating result. The permanent-split-capacitor motor is always a compromise in design between the high-value microfarads needed for starting and the low-value ones needed for balanced running. However, the presence of a flux system from the auxiliary circuit operating in the air gap can greatly reduce the pulsating flux of double frequency, which is the curse of all single-

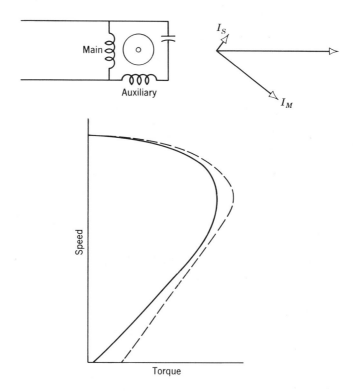

Figure 7-12. Speed-torque curve of a permanent split-capacitor motor. The solid line shows the theoretical torque values of the main winding only (no starting torque). The dotted line indicates the actual torque with both windings in the circuit.

phase motors. Hence this motor can operate comparatively quietly and smoothly.

7-8   Broadened Applications

At one time permament-split-capacitor motors were considered useful mostly for driving fans and blowers, where the starting-torque requirement was low, but smooth, quiet operation was essential.

In more recent times these motors are used on various business machines with a rapidly repeating starting cycle and a very short running cycle, e.g., some types of desk calculators or a ticket-dispensing machine. The operator presses a key, the motor starts, does its function and stops before it is hardly up to speed. Comparatively large capacitors are needed for the starting torque; this results in unsatisfactory running performance, but the motor barely runs and, hence, is not handicapped.

Such frequent starting would blow out electrolytic capacitors and give short life to a centrifugal switch or relay.

Thus, from a beginning of fan and blower applications only, the permanent-split-capacitor motor of special design has become useful in a variety of applications.

7-9   The Capacitor-Start–Capacitor-Run Motor: A Compromise

If we could use several hundred microfarads of electrolytic capacitors for starting, and then remove these from the circuit and add, say, 10 μf of oil capacitors for running, we would have a single-phase motor with excellent starting torque and good running performance. This is especially feasible in motors above 1 hp.

Figure 7-13 shows how the centrifugal switch and capacitors can be connected in the circuit to accomplish the foregoing.

Table 7-1 shows the comparison between a capacitor-start motor and one which is so designed that 10 μf are left connected in its auxiliary winding circuit.

Note that the help of the auxiliary circuit, when running, added to the maximum torque, 10.7 versus 9.8 lb-ft. It increased the full-load speed from 1720 to 1735 (which may or may not be important). But most important is the fact that the main winding did not have to supply all of the power for the load. With the help of the auxiliary winding the current in the main winding reduced from 16.5 to 14.4, and the loss reduced from 102 to 78 w. This may not seem large, but the temperature differences in the two motors was 13°C. That is, as a straight capaci-

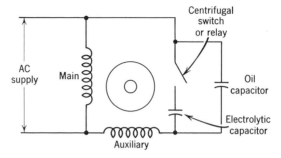

Figure 7-13. Arrangement of a capacitor-start capacitor run motor. The electrolytic capacitor provides high starting torque, the oil capacitor improves running performance.

Table 7-1   A 1½-hp 1,725-rpm Motor

	Capacitor-Start	Capacitor-Start–Capacitor-Run
Main winding	Same	Same
Auxiliary winding		Modified
Running capacitor	None	10 μf
Starting capacitor	540	495
Starting torque	12.5 lb-ft	13.8 lb-ft
Full-load torque	4.57 lb-ft	4.57 lb-ft
Full-load speed	1,720	1,735
Efficiency at full load	70%	77%
Loss in main winding at full load	102 w	78 w
Current in main winding	16.5	14.4
Maximum torque	9.8 lb-ft	10.7 lb-ft

tor-start motor, the main winding running under full load was 13°C hotter than when both windings were operative.

This could mean the difference between an acceptable motor and one that is not. It is especially important as more horsepower is being crowded into smaller and smaller frame sizes.

7-10   The Shaded-Pole Motor

The shaded-pole motor is the simplest and least expensive of all single-phase motors. It is limited, however, in the horsepower ratings available.

Common ratings are from $\frac{1}{10}$ hp (two-pole) down to $\frac{1}{250}$ hp.[1] The efficiency is quite low.

7-11   Construction

A shaded-pole motor is rarely built with distributed windings. Each pole is wound with one coil, as shown in Figure 7-14. In addition, a

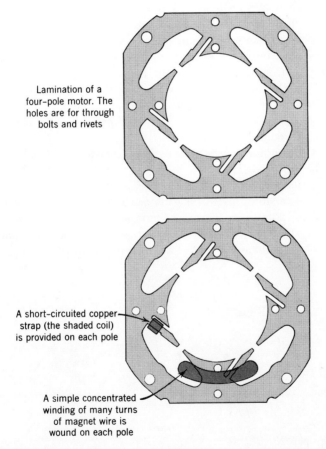

Lamination of a four-pole motor. The holes are for through bolts and rivets

A short-circuited copper strap (the shaded coil) is provided on each pole

A simple concentrated winding of many turns of magnet wire is wound on each pole

Figure 7-14. Stator construction and windings of a four-pole shaded pole motor.

[1] The largest shaded-pole motor of which the author is aware is $\frac{1}{4}$-hp two-pole motor used on bench grinders. The built-in blower required to keep it cool is as large as the motor.

single loop of copper strap is embedded in a portion of each pole face. This is the shading coil.

7-12   Operation

Refer to the two-pole motor shown in Figure 7-15. Consider the instant at which the flux through the whole magnetic circuit is just increasing from zero to some positive value. As this flux builds up through the short-circuit shading coil, it induces a voltage in it, causing a current to flow. We know from Chapter 3 that such a current always builds up a flux which opposes the flux that caused it. The result is that, although the flux of the unshaded portion of the pole builds up as expected, that of the shaded portion builds up an instant later. There is, therefore, a shift of flux from the unshaded to the shaded portion of the pole face. Slight as this is, and a far cry from the ideal rotating flux of the polyphase motor, it is sufficient to start the motor. Of course, this action occurs on each pole, and with both increases and decreases in flux as it goes through its cycle.

Note that this motor always rotates from unshaded to shaded portion. It cannot be reversed. However, in recent years the large demand for window fans that can either exhaust air or draw it in has made it feasible to develop a reversible shaded-pole motor. Note the lamination drawing of Figure 7-16. Now in place of a single short-circuited one-turn coil, a coil of, say, 10 turns of wire is inserted in the slot at each pole edge, as shown. These coils are all connected together into two groups. Short-

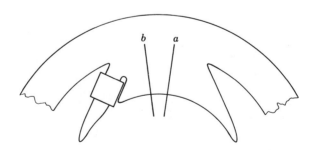

Figure 7-15. Portion of a two-pole lamination. As the flux changes, the short-circuited shading coil chokes off or delays the flux through its portion of the pole, so that the flux of the pole centers at *a*. As the flux magnitude becomes constant, the shaded coil effect is negligible and the flux centers at *b*. This slight flux movement produces the starting torque. As shown, the motor always runs counterclockwise.

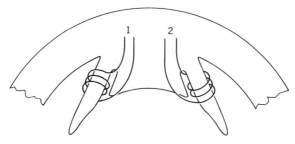

Figure 7-16. Arrangment for reversing a shaded pole motor. Each pole is provided with coils of few turns at the pole edges, as well as the main winding. If all of the coils similar to (1) are connected together and short-circuited, the motor will start and run counterclockwise. Opening the (1) coils and closing those similar to (2) will reverse the rotation.

circuit one group and the motor runs in one direction. Open that set of coils and short-circuit the other group and the motor reverses.

When running, there is always a loss in the shading coil, whether it is of the single- or multiturn type. Also the distribution of the flux in the gap is so irregular that torque pulsations are severe and the motor displays considerable vibration and noise. Nevertheless it fills an extremely useful place in the family of single-phase induction motors.

7-13   Mathematics of the Starting Torque

An examination of the formula by which starting torque for split-phase and capacitor-start motors can be calculated is very enlightening, even though it will not be derived here.

$$\text{starting torque, oz-ft} = \frac{225}{\text{synchronous rpm}} I_m (I_s \sin \theta_t) \frac{R_2}{a}, \quad (7\text{-}1)$$

where I_m = current in the main winding,

I_s = current in the auxiliary winding,

θ_t = angle between I_m and I_s,

R_2 = resistance of the rotor cage,

a = ratio $\dfrac{\text{winding turns in main}}{\text{winding turns in auxiliary}}.$

The expression $I_s \sin \theta_t$ is of special interest. Refer to Figure 7-3 pertaining to the split-phase motor. Although the current in the auxiliary winding is I_s, we know from Chapter 3 that this can be represented

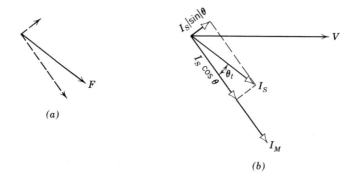

(a)

(b)

Figure 7-17. By separating the locked rotor current of the auxiliary winding into two right-angle components, one can be shown to lead the main winding by 90° as in a capacitor motor. This component is rarely more than 0.5 of the start winding current. (*a*) A force F can be separated into two components, each 90° apart. (*b*) The starting current of the split-phase motor I_S has a component $I_S \sin \theta$ exactly 90° out of phase with I_M.

as two components at right angles to each other. We also know that the ratios of each of these components to the total can be represented as sine and cosine functions. That part of the current which is 90° from the main current is useful in producing torque. This portion, $I_s \sin \theta_t$, can be seen to correspond with the phasor diagram (Figure 7-9), drawn for the capacitor-start motor.

It has been mentioned that θ_t can rarely be increased over 30° in

Table 7-2   Torque and Horsepower Comparisons of Fractional-Horsepower Motors

Type of Motor	Starting Torque	Max. Torque	Full-Load Speeds	Horsepower Range
General-purpose split-phase	Medium	Medium	3,450, 1,725, 1,140, 850	$\frac{1}{20}-\frac{1}{2}$
Special-purpose split-phase	High	High	1,725, 1,140	$\frac{1}{6}-\frac{1}{2}$
Capacitor-start	Very high	High	3,450, 1,725, 1,140, 850	$\frac{1}{8}-1$
Shaded-pole	Low	Low	3,100, 1,550, 1,050, 800	$\frac{1}{250}-\frac{1}{10}$
Permanent-split-capacitor	Low	Medium	3,250, 1,625, 1,075	$\frac{1}{40}-1$
Polyphase	Very high	Very high	3,450, 1,725, 1,140, 850	$\frac{1}{6}-1$

split-phase motors. The trigonometric tables in the Appendix (very abbreviated) show that sin 30° equals 0.5. Immediately we see, then, that our torque formula has a reducing factor of 50%.

We might increase the starting torque by increasing the values of I_s and I_m. To do this, the windings should be reduced in number of turns so as to reduce the impedance. But the sum of I_s and I_m represents the current taken from the line at the instant of starting. National Standards limit such values, with the idea of protecting the motor user from annoying dips in the voltage when his household appliances are started; for instance, a $\frac{1}{4}$-hp motor is not expected to take more than 26 amp when started on a 115-v line (see Chapter 13). Exceptions are made for home laundry equipment on the theory that it is less frequently started than other household devices (ignoring the family of six children). It is permitted to take up to 50 amp. Hence, depending on the application, the starting-current limitations of the split-phase motor become a limit on the starting torque developed.

Now consider R_2, the rotor-cage resistance. An increase in this resistance by using high-resistance alloy or smaller bar and end ring sections

Figure 7-18. A typical single-phase motor rigid base and dripproof construction. The top cover houses the capacitor. (Courtesy of Robbins & Myers, Inc.)

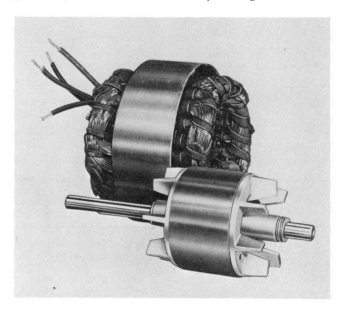

Figure 7-19. This set of single-phase motor parts is sold in sizes ranging from sub-fractional hosepower up to about 10 hp to be built into a motor-driven device. The starting relay and the capacitors (if used) are connected remotely from the motor. (Courtesy of Robbins & Myers, Inc.)

will increase the torque. This can go so far that it reduces the current and, hence, yields no increase in torque. Besides, as we found in the polyphase motor, higher R_2 means reduced full-load speed, more losses, and heat.

This formula illustrates very clearly the advantage of the capacitor-start motor. The angle between I_m and I_s can become approximately 90°. The sine of 90° is unity and the torque expected can become twice that of a split-phase motor. Besides I_s can be larger because of its favorable angle with I_m, because the two of them do not add up to such a large line current. Figure 7-17 shows this favorable addition.

Thus it can be seen that either the standard of industry or the laws of nature place limitations on the starting torque that can be supplied to the product designer (see Table 7-2).

Two typical single-phase motor designs are shown in Figures 7-18 and 7-19.

Chapter 8

Factors Influencing Induction Motor Costs, Prices, and Applications

8-1 Motor Prices

Although it is important for the large user of electric motors to consider all applicable types for driving his product, the induction motors described in Chapter 7 fit into so many fields that it would be wise at this point to compare them. Prices and costs as well as applications will be considered.

Motor prices do not always seem to follow a logical pattern. One can buy a $\frac{1}{3}$-hp 1725-rpm capacitor-start motor at the corner drug store for about $36. The same motor with some modifications is sold in large quantities to an original equipment manufacturer (called in the trade an OEM account) for about $8.50 to $14.00. This is a matter of quantity buying, but another factor affecting prices is the total national market for any type of motor.

Consider first the case of the motors used for home laundry equipment. Total value of these motors produced per year in the U.S. is over $50,000,000. Naturally these are not all identical. They vary from $\frac{1}{8}$ hp to $\frac{1}{4}$ hp in split-phase or capacitor-start designs. Some are 1725 rpm, others are 1725/1140 rpm. But one OEM may purchase 100,000 motors or more of the same kind, and competition for such business is very keen.

The price structure follows the usual competitive pattern. Motor manufacturer A invested in new automated equipment and labor-saving machinery and was able to produce cheaply enough to reduce his prices by 10 cents. Manufacturer B, noting the trend and wishing to stay in business, does likewise and reduces his prices by 15 cents.[1]

---

[1] The author has been unable to find any direct evidence to support the oft quoted statement that an OEM motor buyer would poison his old mother for 15 cents.

This continues until only a comparatively few motor companies end up with the low-cost high-volume automated lines to produce these motors. They are able to produce, sell, and make enough profit to stay in business with current quantity prices of about $8.50 to $10.00.

At the same time, a slightly different motor of the same rating may be sold in 10,000 lots to another OEM account for, say, $13.00. Production of this unit cannot possibly justify the high degree of automation required for the ultimate in cost saving.

The foregoing case illustrates the extreme. To a lesser degree other types of fairly high production motors reflect the same influences in their prices. Tens of thousands of mechanical modifications and an equal number of electrical performance characteristics are required by the motor buyer. Expensive production lines cannot be set up to wring the last penny out of each.

All of this results in an apparently nonconsistent price structure for motors.

8-2   Costs

By far the largest single items making up the cost of an electric motor are lamination steel and copper. In considering the losses in a motor and the need for using laminated steel, it was pointed out that four or five different grades are available. The cheaper the grade, the higher the core loss.

Consider a $\frac{1}{3}$-hp 1725-rpm single-phase motor. The lamination steel "layed down" might cost[2] as follows:

$2.00—typical high grade, low loss.
$1.75—commonly used lower grade.
$1.01—lowest grade practical.

Bear in mind that this is material cost alone. Naturally, after the steel has been made into laminations, heat-treated, and assembled into stators and rotors, the labor and overhead add to these figures. Note also the 50% saving if the cheaper steel can be used. This will be discussed later.

The copper-magnet wire used in the main and auxiliary windings of this $\frac{1}{3}$-hp motor would cost between $1.90 and $2.00.

Note that without considering the end heads, bearings, shaft, body,

[2] The steady increase in the cost of these steels over recent years has never attracted the attention of the White House.

capacitor, or centrifugal switch in this motor, nor any labor nor overhead, we start off with a cost of $3.00 to $4.00 for raw material of just the basic active parts. This is the same motor which might sell for under $10.00.

8-3 Comparative Costs

The shaded-pole motor is the least expensive motor of any type. Suppose we examine its active components as a basis for comparison.

It needs a stack of laminations, a comparatively inexpensive die-cast rotor, and two, four, or six coils depending on the speed required. These coils are machine-wound in place on the poles.

In contrast, a permanent-split-capacitor motor requires double the number of windings (main and auxiliary), and these windings are distributed in a number of slots. These require more time and labor in inserting and connecting. In addition, a capacitor is required. Logically, this motor is more expensive than the shaded-pole type.

The split-phase motor requires the stator, rotor, and two sets of distributed windings, as in the case of the permanent-split-capacitor type, but in place of a capacitor a starting relay or a centrifugal switch and contact mechanism are required. As a split-phase motor goes up in size, say from $\frac{1}{10}$ to $\frac{1}{2}$ hp or above, the cost of the centrifugal switch and contactor remains constant. As a permanent-split-capacitor motor goes up in size, the microfarads required may increase the cost.

Thus there may be an area in which permanent-split-capacitor motors are inherently more costly than similarly rated split-phase motors.

The capacitor-start motor needs two distributed windings, with the auxiliary winding, undoubtedly using more copper than that of the split-phase motor. It requires a centrifugal switch and contactor as well as an electrolytic capacitor. On the basis of these additional components, the capacitor-start motor per given size should be the most expensive.

The very-low-hp three-phase motor that might be compared with these other types is comparatively rare (see Table 8-1). The distribution and connection of its windings may be slightly more expensive than, say, the capacitor-start motor, but, on the other hand, it needs no centrifugal switch, contractor, nor capacitor. Hence its costs are comparatively low.

If all of these motors were built in identical frame constructions, then their logical grading regarding costs would be as previously described. We have already seen that national production and volume influence production-line facilities, tooling available, etc., and these influence *price*. Prices, therefore, do not always follow the apparent cost pattern.

Table 8-1   Active Parts of Various Motor Types Influencing Cost*

Shaded-Pole Motor	Permanent-Split-Capacitor Motor	Split-Phase Motor	Capacitor-Start Motor	Three-Phase Motor
Simple concentrated winding	Two distributed windings Oil capacitor	Two distributed windings	Two distributed windings Electrolytic capacitor	Three distributed windings
		Centrifugal switch	Centrifugal switch	

* Common to all: stack of stator laminations and die-cast rotor.

8-4   Typical Applications

Circumstances alter the types of motors to be used on many applications, and there is also a considerable fringe area of overlap. But, in general, the following shows the usual applications:

General-purpose split-phase motors: Business machines, oil burners, unit heaters, refrigerators, centrifugal pumps.

Special-purpose split-phase motors: Home laundry and ironing machines, cellar drain pumps.

Capacitor-start motors: Pumps, refrigerators, air compressors, business machines.

Shaded-pole motors: Fans, blowers, small oil burners, hair driers, business machines, record and tape players, with gear reducers, advertising displays, home-barbecue spits, etc.

Permanent-split-capacitor motors: Fans and blowers, business machines.

Polyphase motors: A wide variety of small industrial machines, such as machine tools, air compressors, and pumps.

Chapter 9

The Synchronous Motor Types, Operating Characteristics, and Applications

9-1 Synchronous Motor Types

The synchronous motor was mentioned briefly in developing the theory of the polyphase induction motor. It is the purpose of this chapter to discuss more thoroughly the theory and operating characteristics and to deal with the various types.

From the standpoint of the rotor alone three types are found in production: the dc excited rotor, the reluctance-type rotor, and the hysteresis rotor. Any one of these may be used with a suitably wound stator to be connected to a polyphase or single-phase supply. Single-phase synchronous motors are of the split-phase, capacitor-start, permanent-split-capacitor, or shaded-pole types. Although it is theoretically possible to use any combination of stators and rotors to make a synchronous motor, some are not practicable.

9-2 The DC Excited Rotor

Refer to Figure 9-1 which shows the construction of a two-pole rotor. It is built of the usual steel laminations, plus a heavy copper lamination on each end. The entire assembly is held together by copper rivets as shown. These copper end plates and rivets form end rings and rotor bars, not in the usual form of construction, but still capable of acting like a conventional squirrel cage for induction motor action.

Coils are wound on the rotor as shown in dotted lines. The coil ends are brought out to slip rings, as pictured in Figure 9-2. Stationary

139

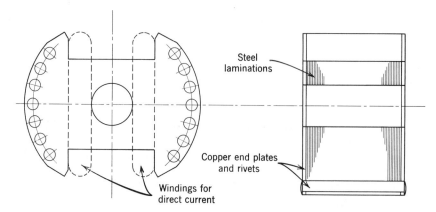

Figure 9-1. Rotor for a two-pole, dc-excited synchronous motor. The copper end plates and rivets form a partial squirrel cage, effective for starting as an induction motor.

brushes bear on these rings to supply direct current to the winding. The rotor is to operate in a conventional three-phase stator.

9-3   Operation

Connect the three-phase lines to a suitable 60-cycle source and the rotor accelerates like an induction motor. The motor reaches a speed of, say, 3550 rpm, more or less, depending on the mechanical load.

Connect the direct current to the rotor winding. At this point the relative speed between the rotating flux of the air gap and the actual

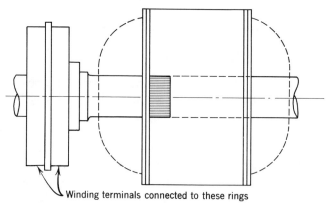

Figure 9-2. Two-pole synchronous rotor with dc winding and slip rings.

speed of the rotor is about 50 rpm. As the rotating flux of the gap passes relatively slowly past the flux system built up by the rotor, the two lock into step. The rotor picks up speed and runs in synchronism. The torque developed to make the final pickup in speed is called the *pull-in torque*. It is important.

9-4   Mechanical Load: Torque Angle

Consider now that a small mechanical load is connected to the shaft of this motor. The motor immediately drops in speed, momentarily, so that the flux of the dc excited field is no longer in exact phase with the rotating flux of the stator. Once the rotor has taken up its new position, the rotor continues to run at synchronous speed, but a few degrees behind its no-load position. This is shown in Figure 9-3. We might consider that the flux lines across the air gap act like rubber bands. Increased mechanical load causes the rotor to drop behind still further, increasing the torque angle α and stretching the bands. Finally, when the angle increases to 90°, the motor pulls out of synchronism. We might say that the rubberbandlike flux lines have now broken. At this point the motor has reached its pull-out torque. Bear in mind that, as we are considering a two-pole motor, the torque angle is actually 90°. If this were a four-pole motor, the angle observed would be 45° mechanically, but still 90 electrical degrees[1] (see Section 2-3).

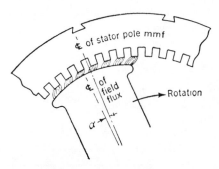

Figure 9-3. A load on the motor causes the rotor to drop back a few degrees so that the voltage applied and the counterinduced voltage in the winding are no longer exactly opposite in position.

[1] Without wishing to go into too much detail, it should, nevertheless, be mentioned that the 90° pull-out torque angle would be exact only if the rotor were a perfect cylinder, with the dc winding distributed uniformly in slots. The fact that in this case so much of the rotor is cut away in one axis to make space for concentrated dc windings actually reduces the torque angle at which pull-out occurs.

With every load change the rotor position will vary, still running synchronously except for the momentary adjustment in position. This is not a simple procedure: the rotor overshoots, comes back, and finally settles down to the correct angle. This can be very troublesome on certain applications, notably tape drives. This momentary oscillation can be reduced by the damping effect of the squirrel cage in the face of the rotor poles or, more successfully, by the use of a flywheel on the shaft.

9-5   Why a Torque Angle is Necessary

We know that whenever flux cuts conductors a voltage is generated. Because this magnetic field (tied in with the rotor structure) rotates past the stator winding, a voltage is generated in each phase. If the motor carries no load and, ideally, has no losses, the generated voltage in each phase is exactly equal and opposite to the applied voltage. Under this state of balance no current would flow in the motor windings. When a mechanical load is connected on the shaft, the motor must draw current to obtain the power for the load. It can only do this by dropping back through the angle α (see Figure 9-4). Now, even though the voltage generated and the voltage applied per phase may be equal, they are no longer 180° out of phase. The resultant voltage E_R is useful in sending current through the winding. Because of reactance, this current lags greatly behind the voltage which causes it, but note from the diagram that it may be nearly in phase with the supply voltage. The power input to the motor is now the product of volts per phase, the in-phase portion of I_{line} and the number of phases. This input minus the losses must be the mechanical power available at the shaft.

Increasing the torque angle α increases E_R and permits more current and power input.

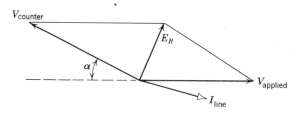

Figure 9-4. If the applied voltage and the reduced (counter) voltage are equal and opposite, the ideal motor will take no current. Loading causes the rotor to drop back through the angle to create a difference between these voltages equal to E_R. This voltage causes the current I line to flow through the winding. $V_{applied}$ and the in-phase portion of I_{line} represent power input.

9-6   Numerical Consideration of Torque Angle: Pull-Out Torque

If we ignore the slight error caused by the lack of magnetic symmetry in the two-pole rotor under consideration, and assume that the torque developed varies as the sine of the torque angle, we can obtain a better understanding of the effect of torque-angle requirements on motor size.

Consider a case in which the user required a synchronous motor to operate from no load to full load with a change in angle of only 10°. This was fixed by the type of equipment which was to be driven. The sine of 10° (see Appendix) is 0.173. This means that the full-load torque is 17.3% of the maximum or, conversely, the maximum torque is 580% of full load.

In the normal design of a synchronous motor the maximum torque may be 200 to 250% of full load. It follows, then, that this motor will be at least twice as large as one designed for more normal characteristics.[2]

9-7   Power Factor Control

A dc-excited synchronous motor possesses a peculiar quality. For any given load it can be made to draw current with either a leading or lagging power factor. In Figure 9-4 we saw that the motor could draw current only because the counterelectromotive force (voltage induced in the winding) was no longer directly opposed to the applied volts per phase. Suppose we increase the voltage applied to the dc winding. This increases the direct current, the flux, and, hence, the induced emf. Now the difference in voltage E_R is not only caused by the phase difference but also by an actual difference in magnitude of the voltages (see Figure 9-5a). The position of E_R is such that the resulting line current, although still lagging approximately 90° behind it, nevertheless leads the applied voltage. The motor draws leading current like a capacitor circuit. The true power input need not change, again it is the product of line voltage and the in-phase portion of the current.

In a similar manner, reducing the voltage applied to the dc winding

[2] If the author appears to belabor the importance of torque angle and its necessity, it is because, in dealing with project engineers who use synchronous motors, the theory seems difficult to put over. Projects frequently start with the statement from the prospect that he needs a motor that stays absolutely constant in speed and never drops back or varies instantaneously from synchronism! Repeal Newton's third law.

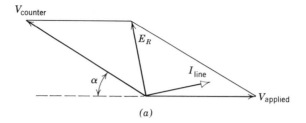

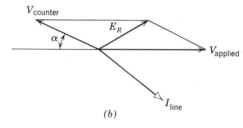

Figure 9-5. Varying the dc$_\text{applied}$ to the rotor winding increases or decreases the counter electromotive force. The current lags the resulting voltage, E_R by about 90° because of the inductive reactance of the stator winding. This current may lead or lag the applied voltage as shown, a characteristic true only of excited synchronous motors. (a) Motor takes a leading current. (b) Current now lags the applied voltage.

reduces the counterelectromotive force and swings E_R to the position shown in Figure 9-5b. Now the line current lags the applied voltage.

This quality is made use of only in large synchronous motors which may be purchased to operate at unity power factor or at 0.8 leading power factor.

9-8 Applications

The dc-excited synchronous motor just described is most frequently used in large integral-horsepower sizes. Pumps, compressors, and other large industrial equipment are frequently driven by slow-speed synchronous motors rated as high as 2000 hp. They offer the advantage of high or leading power factor, as just described, helping to neutralize the lagging current taken by the majority of industrial motors. Thus one factory equipped with hundreds of induction motors, all taking a lagging current, may be helped by several large synchronous motors taking a leading current. Improvement in overall plant power factor is sometimes rewarded by the power supplier through reduced rates.

In addition, small dc-excited synchronous motors are sometimes built in fractional-horsepower sizes for certain types of precision drive. Early television work, radar equipment, and tape drives were some examples. In general, we can obtain more power in a given frame from this type of motor than from any other of the synchronous types. These motors are relatively expensive.

9-9   Reluctance-Type Rotors

Examine the two-pole rotor of Figure 9-1. Although the large cutaway portion was provided for a dc winding, this rotor would, nevertheless, operate at synchronous speed without such a winding. Just as current seeks the paths of least resistance so does the magnetic flux set up in the path of least reluctance. If this rotor were placed in a suitable two-pole field, it would accelerate as an induction motor by virtue of its partial squirrel cage. If the inertia of the load were not too great, it would pull into synchronism with the rotating flux tied in through the longitudinal length of the poles. We might say that the flux prefers this path to that at right angles, which involves a much greater length of air gap. We thus have a reluctance-type synchronous motor (see Figure 9-6).

Note the rotor lamination of Figure 9-7. Here four sections of the periphery are cut away for operation in a four-pole field. The slots

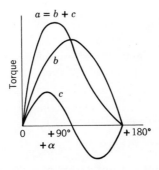

Figure 9-6. Torque angle in electrical degrees for a synchronous motor with a symmetrical, direct-current excited rotor is shown by curve b. If the rotor surface is partly cut away, as in Figure 9-1, the torque characteristic follows curve a. A reluctance-type rotor with no winding acts as shown in c.

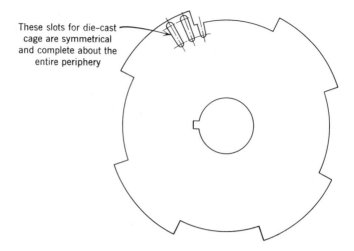

These slots for die–cast
cage are symmetrical
and complete about the
entire periphery

Figure 9-7. An ordinary squirrel-cage induction motor rotor can be made into a four-pole reluctance-type synchronous motor with the four large notches shown in the periphery.

for the rotor bars, of which only a few are shown, actually continue around the entire circumference. A stack of these laminations is built up, fitted into a mold, and injected with molten aluminum, exactly as in the case of the conventional die-cast aluminum squirrel cage. Now the four outer sectors fill with aluminum also. It is easier to permit this than to attempt to block them off. Being nonmagnetic, these aluminum sectors have no effect on the relative reluctance paths for the flux.

9-10   Operation

Loading this type of motor has practically the same effects as those described for the dc-excited case. It drops back momentarily in speed, through a torque angle α. The more load, the greater the angle. However, this type of motor reaches its maximum torque when it has dropped back 45 electrical degrees (see Figure 9-8).

9-11   Other Rotor Configurations

A much more sophisticated type of reluctance rotor is shown in Figure 9-9. This also is for four-pole operation. Imagine the flux from the air

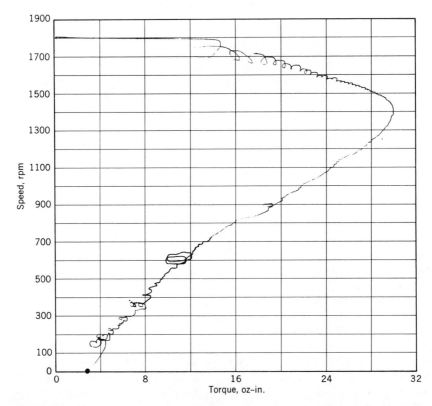

Figure 9-8. An automatic device called an *x-y* plotter is capable of drawing the speed-torque curve of a motor on a chart, which represents a curve for a reluctance-type, permanent-split capacitor, synchronous motor. The full-load torque of this motor is about 7 oz-in. It would pull out of synchronism at about 14 oz-in. The irregularities in the lower part of the curve represent changes in torque as the rotor teeth approach or pass the stator teeth. The loops in the upper part of the curve represent varying pull in torque close to synchronous speed as the revolving field flux approaches and then passes the reluctance poles.

gap running parallel to the curve slots. This is the position of minimum reluctance. Now assume that we moved this flux system 45°. The flux through one pair of stator poles would have to cross the air gaps *aa* in the rotor, and would encounter a comparatively great increase in reluctance. Well-designed rotors of this type make it possible to obtain almost the same output for a given amount of active material as we would expect from an induction motor.

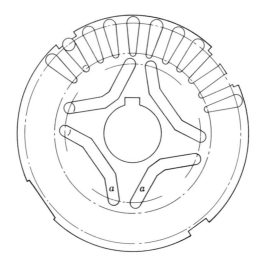

Figure 9-9. Greater reluctance torque can be obtained from this synchronous rotor. Minimum reluctance is obtained when the four-pole air-gap flux lines up parallel with the cutaway sectors. Greatly increased reluctance would be presented to the flux if it had to cross the sectors as at *a-a*.

9-12 Hysteresis Rotors

In dealing with the losses in steel caused by an alternating flux, mention was made of eddy currents and hysteresis. In the latter case the reversal of the flux was considered causing internal molecular friction which appeared as a heat loss. It is now necessary to consider hysteresis in a more thorough manner.

Refer to Figure 9-10. A bar of special alloy steel is surrounded by a coil through which a controlled direct current flows. This current can be varied and reversed, as shown in the circuit diagram. Suppose that a certain current is flowing in the coil and the flux in the steel is up to point *a*. Now reduce the current. When it reaches zero, the steel is still magnetized to point *b*. Next reverse the current. A "negative" current is necessary to reduce the flux to zero, as shown at point *c*. Continue increasing the current (in the negative sense) until the flux is now of *d* magnitude. Reduce the current and the flux reduces to *e*. Reversing the current and increasing it gradually carries the flux along the line *ea*.

The point is that, once steel is magnetized in one direction, it tends to retain that magnetism. Even reducing the current (and the ampere-turns, NI) to zero, leaves a flux of *ob* in the bar. A negative magnetic force of *oc* was needed to reduce the flux to zero.

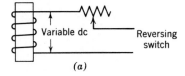

(a)

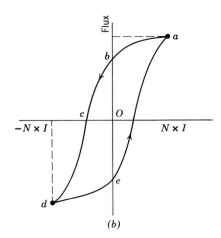

(b)

Figure 9-10. Gradually magnetizing and demagnetizing a bar of steel shows that it tends to retain its magnetic flux to the magnitude of Ob or Oe, even when the magnetizing force (the current in the coil) reduces to zero.

This figure is called a *hysteresis loop*. It varies greatly for different types of steel. (Hysteresis means, very freely translated, effect lagging behind the cause.)

If we take any induction motor, polyphase or single-phase, remove its conventional rotor, and place a cylinder of suitable alloy steel inside the stator to form a new rotor, connection of the stator winding to a suitable voltage source will cause the rotor to pick up speed and run in synchronism. In a sense, this cylinder of steel supplies its own squirrel cage by having induced currents flow in various and complex paths through its surface. Hence starting and accelerating torques are all caused by eddy currents. Synchronous pull-out torque is caused by the ability of the steel to stay magnetized. The maximum torque is a measure of the area of the hysteresis loop.

These motors are built in many mechanical forms. A typical rotor construction is shown in Figure 9-11.

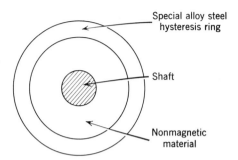

Figure 9-11. Cross section of a typical hysteresis-type rotor.

9-13 Applications

In the very small millihorsepower sizes hysteresis motors can be built of simple stampings and can be produced cheaply. Clock motors or other timing devices are operated by this means.

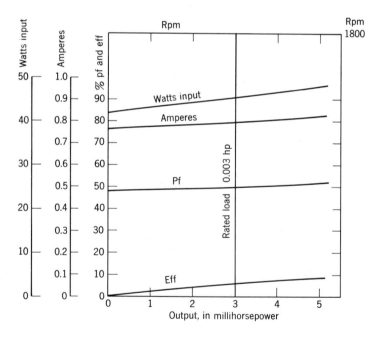

Figure 9-12. Typical performance of a reluctance-type rotor and a four-pole, shaded-pole field. Note the low efficiency. Full-load torque = 1.66 oz-in. Break-down or pullout torque = 3.72 oz-in. Starting torque = 3.6 oz-in. Locked rotor amperes = 1.19 at 115 v.

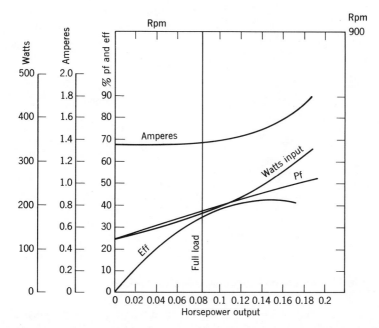

Figure 9-13. Typical performance of a three-phase synchronous motor with an eight-pole reluctance-type rotor, rated at $\frac{1}{12}$ hp. Full-load torque = 7.78 oz-ft. Breakdown torque = 15.4 oz-ft. Starting torque = 47.8 oz-ft. Locked rotor amperes = 3.23 at 208 v.

As the sizes increase, the cost goes up fairly rapidly, for two reasons. The alloy steel used in the rotor is expensive. If we take any reluctance-type synchronous motor and replace the rotor by a suitable hysteresis construction, it will be noted that the watts go up and the maximum torque goes down. It becomes a weaker, less efficient, motor. Hence for a given rating a hysteresis motor is likely to be larger than the reluctance motors (see Figure 9-12).

The rotor is completely symmetrical. Hence the same rotor will operate in a two-, four-, six-, or eight-pole field. For various tape drives and timing devices, motor speeds of, say, 1800 and 1200 rpm might be desirable. By providing two sets of windings on the stator (four- and six-pole) and switching from one to the other, these speeds can be obtained.

This is impossible with reluctance-type rotors. However, the configuration of the reluctance rotor can be made to operate at speeds in a 2 to 1 (integral) ratio, e.g., 3600 and 1800 rpm (see Figure 9-13). The available horsepower varies with the speed.

Chapter 10

Direct-Current Motors, Operating Characteristics, Electronic Supply

10-1   Review of Elementary Motor Action

We have already seen in Chapter 1 and again in considering all of the induction motor operations that a conductor carrying current in a magnetic field builds up a flux which reacts with the field and produces torque. Of course, the physical structure of the dc main field is stationary and its flux is stationary in space and time.

In Chapter 1 we dealt with force on a single conductor which moved it at right angles to the main flux to the outer boundary of the field at which the force reduced to zero. In a dc motor we will find that some means is required for reversing the armature current so that torque is exerted through a complete rotation. The device used is called a *commutator*.

10-2   Motor Action with A Commutator

Figure 10-1 shows direct current being fed to an armature of a single-loop coil. The reaction of the armature flux and the field flux produces a torque which varies from position *a* through *b* to *c* until, finally, at point *c* the torque reduces to zero and this elementary motor would stop there after making one-half revolution. It is at this point that we must reverse the current in the armature loop by using a two-part commutator rotating with the armature. This is shown in Figure 10-2 which indicates that we now have a motor which will run, but its torque is zero at two points in its rotation.

152

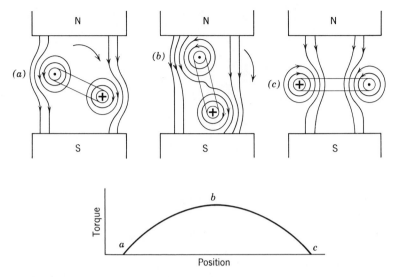

Figure 10-1. A current-carrying loop in a magnetic field builds up torque through one-half revolution. To obtain a complete revolution the current in the loop must be reversed at position (c) and again just ahead of (a). Only a portion of the field flux is shown. It is assumed to be uniformly distributed over the pole faces.

Figure 10-3 shows this first loop duplicated and connected to two opposite segments of a four-part commutator. Now the torque never goes to zero. When the first coil is midway between the poles, it is disconnected from the dc armature supply, producing no torque; but at that time the second coil is connected and produces a maximum value. We can see here the comparison of two-phase alternating current, but now there are no negative loops because of the commutator.

Figure 10-4 shows a closed-circuit winding in which all coils are connected. Each set of coils aids in producing torque at various positions.

10-3 Conventional Armature and Windings

The usual armature lamination is made of sheet steel of various grades and has slots around its periphery with rather large openings for automatic winding insertion. We have seen in the previous illustration the advantages of having multiple coils so connected that all of the coils are useful.

Figure 10-5 illustrates a conventional winding classified as *lap, two-*

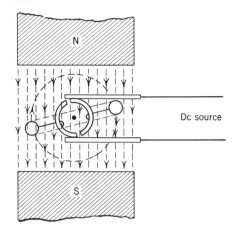

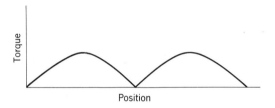

Figure 10-2. The single-loop armature winding now has its ends connected to a two-part commutator on which long contacts are used to make connection to the external source. At the position shown in only a few degrees of rotation the current will be reversed in the coil and the current and torque will be as shown in Figure 10-1*a*, ready to complete its rotation.

layer. As for a four-pole motor, four brushes are used and the input lines connect to each *set* of brushes. Note that the coil identified as 1-12 is short-circuited by a brush at the instant shown. Such instantaneous short circuits occur at each brush as the motor rotates (see Figure 10-6).

A second type of armature winding is shown in Figure 10-7. This, again, is double-layer winding, but it is a "wave" rather than a "lap" winding. Notice that the end of any coil does not connect to a commutator segment adjacent to the beginning of that coil, as was done with the lap winding. Consider the coil connected to commutator segment 1, progressing to the right to connect to segment 13. That coil is between the poles and is inactive. Hence we could eliminate the bottom brush

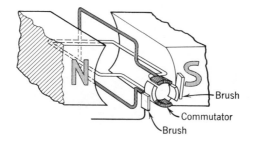

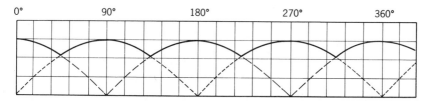

Brush

Commutator

Brush

Figure 10-3. With a four-segment commutator and two coils at right angles, the torque never goes to zero.

on this commutator because coil 1-14 would fulfill the same function as the cross connection between these two brushes. Tracing this through with other coils and brushes shows that, even though this is a four-pole motor, two brushes 90° apart would operate the motor as successfully as the four brushes shown.

This point is brought out because a novice is sometimes troubled in finding a four-pole dc motor having only two sets of brushes. He can then be sure that the armature is wave-wound. In the mechanical construction of the motor and for reasons of accessibility, it may be more

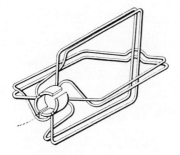

Figure 10-4. An elementary closed-circuit winding. All coils carry current, although there are instants at which any coil produces zero torque.

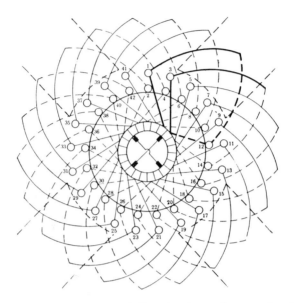

Figure 10-5. A two-layer armature winding for four-pole operation. Input is to each set of diametrically opposite brushes.

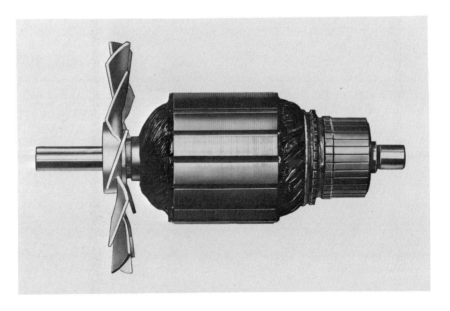

Figure 10-6. Armature assembly for a small high-speed dc motor. The portion of the commutator on which the brushes ride has been smoothly machined. (Courtesy of Robbins & Myers, Inc.)

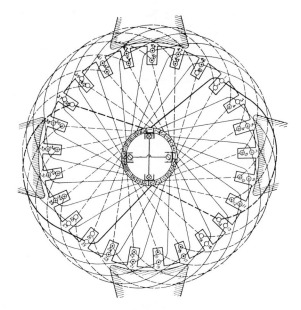

Figure 10-7. A two-layer armature winding of the wave type. This would operate successfully with only two brushes 90° apart.

convenient to use two brushes rather than four. Of course, they must be larger in such a case, because brushes are limited by recommended current-carrying capacity.

10-4 Generator Action: Interchangeability

If the armatures as described were to rotate in a magnetic field, driven by some external means, the coils would have voltages generated in them which would be sufficient for a dc voltage to be collected at the brushes. In fact, every dc motor *is* a dc generator at the same time that it is developing mechanical power at the shaft. Why not? All of these conductors are cutting flux and building up a voltage. *The voltage applied to the brushes as a motor must be higher than that which is generated internally.* The generated voltage is thus a countervoltage or a back emf which uses up the greater part of the voltage applied. Thus

$$V_{\text{applied}} = E_{\text{counter}} + IR_{\text{drop in arm.}} \qquad (10\text{-}1)$$

This is the fundamental action versus reaction equation for the motor. We shall see later how it determines the motor speed.

In the same manner, every dc generator has motor action as soon as an electric load is applied to the brushes. The output current flows through the armature winding in the direction that would produce motor action. The driving machine for the generator must overcome this *backward* torque. It is in this manner that the driver "senses" the magnitude of the electric load on the generator. More current flowing through the load means a greater current and flux system for the armature conductors and, hence, more torque is required to operate the generator.

Now the equation for the voltages becomes

$$V_{\text{output}} = V_{\text{generated}} - IR_{\text{drop in arm.}} \tag{10-2}$$

10-5   Armature Reaction

It has been mentioned that these dc machines will run as either motors or generators. One slight correction is necessary, which involves a shift of the brushes from what might be called a *neutral position*. This is because of *armature reaction*.

Refer to Figure 10-8*a* showing the usual distribution of the field flux for a two-pole machine. For convenience we will represent the armature conductors as being on the armature surface. We assume there is no current in the armature.

Figure 10-8*b* shows the flux system built up by the armature alone when the brushes are in the position shown. (The commutator is not pictured but the effect is the same.)

Figure 10-9 shows both field and armature excited as for normal operation. The two flux systems react, crowding the flux in one pair of pole

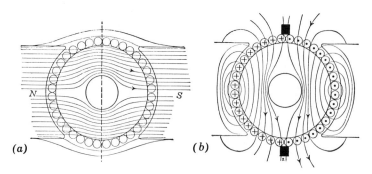

(a)   (b)

Figure 10-8. A two-pole machine showing normal flux position for the field in (*a*) and for the excited armature in (*b*). The two-flux systems will interfere with each other or combine when both armature and field are excited.

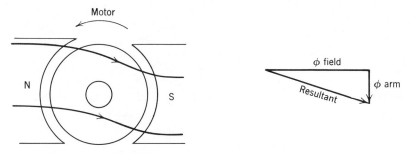

Figure 10-9. The resultant flux is twisted by the action of the armature-magnetizing force on the field. Note in this case that the motor would rotate as shown and the "neutral axis" across the flux would be shifted backward rather than perpendicular to the poles. For most effective operation and best commutation the brushes in a motor should also be shifted back against the direction of rotation. (The use of the fluxes as vectors to show the position of the resultant is not quite correct because of the effect of saturation in the steel which would usually reduce the resultant.)

tips and reducing it in the other tips. The lines of flux are no longer parallel with the poles.

But for best commutation and the most effective use of the conductors, the coils should be commutated so as to be "sliding along" the flux at that instant rather than cutting it. In short, we should shift the brushes (see Figure 10-10).

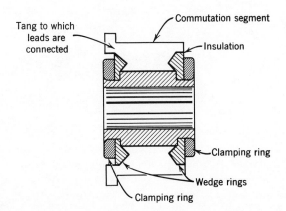

Figure 10-10. Cross section of a typical commutator construction. In some cases these segments are held in place and insulated by molded plastic. When Thomas Davenport invented the dc motor, he first used a cam to open and close armature circuits at suitable points. Later he developed a crude counterpart of the modern commutator.

Analysis shows that the flux has been twisted in the direction of rotation in the case of a generator, but in the opposite direction for motors.

10-6   Reversing Small DC Motors

Some applications require that a motor be operated in either direction of rotation. In such a case the brushes must remain on neutral midway between the poles, with the resulting disadvantages in sparking and poor commutation.

Refer now to Figure 10-11. Note here that two holes are provided in fixed positions in the end head for the cartridge brush holders. The head cannot be shifted. But if, when the motor is built, it is known in which direction the motor will always rotate, the effect of brush shift can be brought about by displacement of the armature windings (see Figure 10-12). Such a motor, once built, should operate in one direction only.

Figure 10-11. Cutaway view of a small dc motor which also shows a section of the cartridge-type brush holder. Note the vulcanized fiber cartridge and the brush with a spring between it and the black bakelite screw cap. Removal of this cap enables the brush and spring to be replaced. Inside the fiber tube is a brass tube with a square hole in which the brush rides loosely. The brass tube projects beyond the fiber and a "garter" spring fits around it, making an electric contact with a lead. Thus the current is fed from the garter spring to the brass tube and brush, then through the commutator to the armature winding. (Courtesy of Robbins & Myers, Inc.)

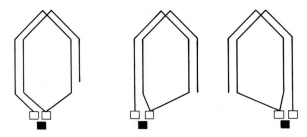

Figure 10-12. Portions of armature winding connected to commutator bars showing how one can obtain the effect of brushes on neutral, or apparent brush shift backward for either direction of rotation. The brushes remain unchanged in their position in relation to the motor end head.

10-7   Reversing Large DC Motors: Interpoles

Larger dc motors, say about 1 hp and over, are usually equipped with a different arrangement for holding the brushes than that already shown. For motors of this size the entire brush stud and brush holder assembly can be turned through a considerable angle. Preparing the motor for most effective operation in either direction involves turning the assembly.

However, if the normal operation of the motor requires that it be

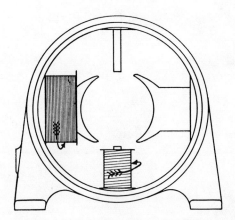

Figure 10-13. Field frame for two pole machine with interpoles. Direction of field currents shown for clockwise rotation of generator armature. Interpole precedes main field pole of like polarity on generator. Interpole follows main field pole of like polarity on motor. (Courtesy of The National Carbon Co.)

run in one direction and then reversed (as a part of the driven-machine cycle), the large motor has the advantage of *interpoles* which compensate for the flux-shifting effect of armature reaction (see Figure 10-13).

The interpole (or commutating pole) has a heavy winding connected in series with the armature. It opposes the magnetizing force which tends to cause flux shift. As the motor reverses, the current in the armature and armature reaction reverses, but so does the magnetizing force of the interpole. As a result, in the ideal case, regardless of direction of rotation, the motor commutates in the neutral plane. It is not always

Figure 10-14. Armature and field of a 10-hp dc motor showing interpoles for improved commutation.

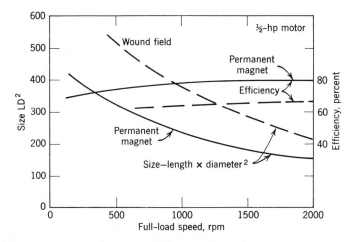

Figure 10-15. Comparison of permanent magnet versus wound field dc motor. Smaller size per horsepower results from the permanent magnets. (Courtesy of *Product Engineering*)

necessary to use as many interpoles as main poles. One interpole per pole pair can give satisfactory results (see Figure 10-14).

10-8   Motor Types: Connections

Up to this point we have been discussing flux in the field and armature current and flux reacting with each other to obtain torque. But how are these two components obtained and how are the windings connected?

Permanent Magnet Fields. Permanent magnets have been used as pole pieces in various small- and medium-sized motors (see Figure 10-15). The heavy steel shell into which the moulded magnets are fastened completes the magnetic field circuit. This has been made practicable because the improvements in permanent magnets result in useful high-flux densities without too much reduction with aging. Direct current is supplied only to the armature. More recently, ratings up to 100 hp are being marketed using permanent magnet fields.

Shunt Motors. The armature and field are connected in parallel (see Figure 10-16a). (Our forefathers spoke of the field being "shunted" across the armature, hence the name.) The operating characteristics are discussed in detail later, but, for the time being, we shall identify this motor as being essentially a constant-speed motor.

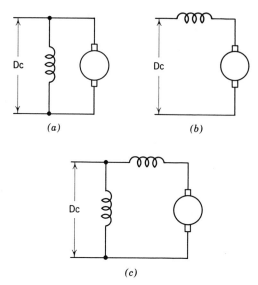

Figure 10-16. Three common types of dc motor, each displaying different operating characteristics. (*a*) Shunt motor. (*b*) Series motor. (*c*) Compound motor.

Series Motors. Here the field of comparatively few turns of heavy wire is connected in series with the armature. Naturally, with such a connection, both windings must carry the load current. This motor has varying-speed characteristics, slowing down with heavy loads.

Compound Motor. This type of motor has the usual shunt winding of many turns and a series winding. Depending upon the method of connection of the series field, adding or subtracting from the magnetizing force of the shunt field, it can cause the motor to drop in speed with load or increase in speed with load (see Figure 10-17).

If the series field is weak, with only a few turns, and it is used chiefly to "stabilize" the speed, this motor is by definition still a *shunt* motor.

10-9 Operating Theory

Consider that a shunt motor is connected to a suitable source of direct current. The current in the field winding builds up almost at once (slowed slightly by the self-inductance, as described in Section 1-13) to a final value determined by the resistance of the winding. The current through the armature circuit is very high, again limited only by the low resistance of the armature winding and the contact drop at the brushes. Because of this heavy current, the starting and accelerating torques are high.

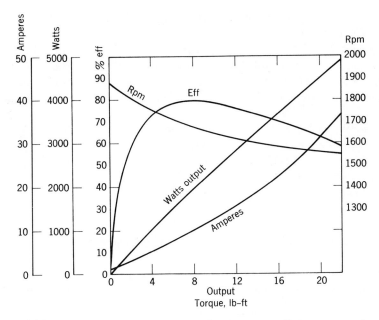

Figure 10-17. Typical operating characteristics of a 3-hp, 230-v compound wound dc motor.

But as the armature gains in speed, its conductors cutting the field flux generate a voltage, previously identified as a counterelectromotive force. At all times during this acceleration,

$$V_{\text{applied}} = E_{\text{counter}} + IR_{\text{drop in arm.}},$$

as shown in Eq. 10-1.

The counterelectromotive force, as with any generated voltage, depends on the speed of cutting flux. Hence

$$E_{\text{counter}} \propto \text{rpm} \times \text{field flux.}$$

During the acceleration the current in the armature circuit decreases as the counterelectromotive force builds up. A final speed is reached when the counterelectromotive force is practically equal to the applied voltage, allowing only enough current in the armature to produce the torque needed to overcome the friction and windage losses plus any core losses in the laminated-steel armature.

Now, if a mechanical load is connected to the shaft, the motor slows down. The counterelectromotive force is reduced and the armature draws more current until its interaction with the field produces just enough torque to handle the connected load. Increased load slows down the motor still further. It might be said that the counterelectromotive force acts as a "valve" or governor, fixing the input current and speed to match the mechanical load.

The shunt motor as described displays a drooping speed-load curve, but other factors may interfere. Without interpoles, increased armature current could distort the field to an increasing amount, saturating the pole tips so that the net field flux might be reduced and the motor speed up.

Normally with a reduction in flux, the motor must run faster to build up its counterelectromotive force. Also, as the motor continues to operate, the field winding heats up, giving it a slightly higher resistance. This further reduces the flux and speed. Conversely, increase in armature resistance with heating *reduces* the speed.

10-10   Standards for Fractional Horsepower Motors

Fractional-horsepower dc motors are expected to be available in speeds corresponding to 60-cycle induction motors, namely 3450, 1725, 1140, and 850 rpm.

Of these, National Standards recommend (see Chapter 13 concerning NEMA) that the variation in speed based on full-load speed should not vary more than the following percentage. A constant temperature is assumed.

As to the effect of heating, these same standards recommend:

"The variation in speed of direct-current motors (built in frame 42 and larger) from full-load cold to full-load hot, during a run of a specified duration and based on the full-load speed hot, shall not exceed 15 percent for enclosed motors or 10 percent for all other types of motors."

Table 10-1

Horsepower	Speed (rpm)	Shunt-Wound (%)	Compound-Wound (%)
$\frac{1}{20}-\frac{1}{8}$, incl	1725	20	30
$\frac{1}{20}-\frac{1}{8}$, incl	1140	25	35
$\frac{1}{6}-\frac{1}{3}$, incl	1725	15	25
$\frac{1}{6}-\frac{1}{3}$, incl	1140	20	30
$\frac{1}{2}-\frac{3}{4}$, incl	1725	12	22
$\frac{1}{2}$	1140	15	25

10-11   Adjustable-Speed and Adjustable Varying-Speed Motors

It should be obvious now that in the dc shunt motor we have a very simple and comparatively inexpensive device for obtaining an adjustable-speed motor. A variable resistance connected in series with the shunt field can weaken the field flux and increase the speed. With the resistance set at various values, speed versus torque curves result, as shown in Figure 10-18. All speeds so obtained are above the base speed. Because

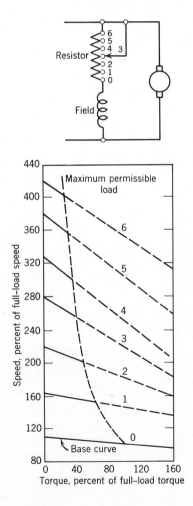

Figure 10-18. Operation of a shunt motor above base speeds by shunt field control. (Courtesy of *Machine Design,* Motor Reference Issue.)

the field current is small, the current capacity of this added resistance need not be high.

If we are going to apply an adjustable-speed shunt motor to a load requiring a great variation in speed, the natural question arises as to how high a speed can be expected. Is more or less horsepower available at different speeds? These values are set by National Standards, which also cover various degrees of overload for various ratings, which supplement Table 10-2, but are omitted.

The addition of variable resistance in series with the armature circuit reduces the speed. (Because of this added IR drop, the motor needs less counterelectromotive force to balance the applied voltage.) Such

Table 10-2

Horse-power at Base Speed	Horsepower at 300 Percent of Base Speed and Higher Speeds (dripproof motors only)	Base Speed (rpm)									
		3500	2500	1750	1150	850	650	500	400	300	
		Speed by Field Control (rpm)									
½	0.65	....	....	....	....	3000	2600	2000	1600	....	
¾	1	....	....	....	3200	3000	2600	2000	1600	....	120 and
1	1.3	....	....	3500	3200	2800	2600	2000	1600	....	240
1½	2	4000	4000	3500	3000	2800	2600	2000	1600	....	
2	2.6	4000	4000	3300	3000	2600	2600	2000	1600	1200	
3	4	4000	3700	3300	2800	2600	2600	2000	1600	1200	
5	6.5	3700	3700	3000	2800	2600	2400	2000	1600	1200	
7½	10	3500	3500	3000	2800	2600	2400	2000	1600	1200	
10	13	3500	3500	3000	2800	2500	2200	2000	1600	1200	
15	20	3500	3300	3000	2600	2500	2200	2000	1600	1200	
20	26	3500	3300	3000	2600	2400	2200	1800	1600	1200	
25	33		3100	3000	2600	2400	2000	1800	1600	1200	240 v
30	40		3100	3000	2600	2400	2000	1800	1600	1200	
40	52		3100	2700	2400	2200	2000	1800	1600	1200	
50	65			2700	2400	2200	1800	1800	1600	1200	
60	80			2400	2200	2000	1800	1600	1600	1200	
75	100			2400	2200	2000	1800	1600	1600	1200	
100	130			2200	2000	1800	1600	1600	1600	1200	
125	165			2000	2000	1800	1600	1600	1600	1200	
150	200			2000	2000	1800	1600	1600	1600	1200	
200	260			1900	1800	1700	1600	1600	1200	1200	
250					1700	1600	1600	1400	1200		
300					1600	1500	1500	1300	1200		250 and
400					1500	1500	1400				500 v
500					1500	1400					
600					1500	1300					500 and
700					1300						700 v
800					1250						

From NEMA, MG 7-10-61.

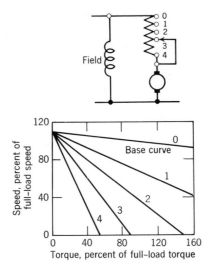

Figure 10-19. A shunt motor with variable resistance in the armature circuit gives speeds below the base speed and reduce more rapidly with load increase. (Courtesy of *Machine Design*, Motor Reference Issue.)

resistances must be of heavy current capacity, matching that of the load current. The I^2R loss in them, which can be large if the speed is greatly reduced, detracts from the efficiency of the motor. This resistance at various settings, gives a series of speed versus horsepower or torque curves below the base speed (see Figure 10-19).

Light loads require smaller armature current and the IR drop is then comparatively small. Heavy loads result in larger IR drops; hence the speed reduces with the load. This fulfills the definition of *varying speed* resulting from the load rather than from the operator's control; but as the resistance can be varied by the operator, this is then an *adjustable varying-speed* motor.

10-12   Reversibility

It should be clear that reversing the direction of the current in either the field or the armature builds up torque in the opposite direction and reverses the motor. Reversing the motor leads as they connect to the line does not reverse the motor, because the double reversal still results in torque in the original direction.

The field circuit should never be opened when the armature is connected. With the flux reduced to nearly zero (except for a slight residual magnetizm in the steel), the armature rotates at extreme high speeds in attempting to build up the required emf. In very small motors friction or ventilating fan loads may hold this speed down below a destructive value. But in most fractional and in almost all larger horsepower motors, the commutator may fly apart and the armature windings push out through the slots. The degree of destruction is rather amazing.

10-13   The Series Motor

When a series motor is connected to the line, the input current is high and so is the field flux, limited somewhat by magnetic saturation. The resulting accelerating torque is large, and the motor quickly obtains nearly a runaway speed, perhaps six times the normal load speed in the case of the smaller motors. Here the field flux does not remain constant as in the case of the shunt motor. Low input current at light or no load means low field flux, and the armature must rotate at high speeds to build up sufficient counterelectromotive force. Conversely, an increased load results in both a higher field and armature current, and the resulting large field flux requires a slower speed to build up the counterelectromotive force. Hence the speed changes greatly with load (see Figure 10-20). Speed control can be achieved as shown in Figure 10-21.

The reduction in speed with load, inherent with this type of motor, has given rise to the general term of *series characteristic* (see Figure

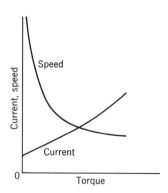

Figure 10-20. Typical speed-torque curve for the dc-series motor. The no-load speed may be about six times the rated value.

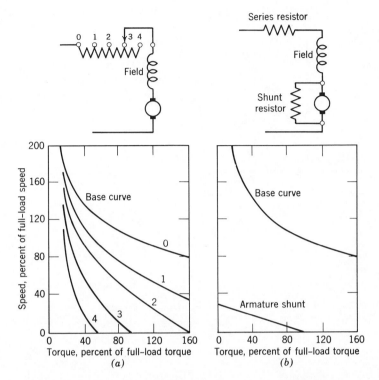

Figure 10-21. Speed control of a series motor can be obtained by a variable resistance in series with the line, as in (a). This really reduces the voltage applied to the motor. A combination of series resistor and a second resistor in parallel with the armature will also reduce the speed (b). The parallel resistor reduces the armature current, leaving a normal field current, naturally resulting in a reduced speed. (Courtesy of *Machine Design*, Motor Reference Issue.)

10-22). Thus, if a special induction motor is built so that its speed drops off rapidly with load (extremely high slip), we speak of the motor as having series characteristics.

10-14 Braking

With sufficient control equipment it is possible to bring the dc motor down to a stop from its normal operating speed, reverse it, and bring it up to speed again in the opposite direction. In general, this can be done more smoothly with dc motors than with ac ones.

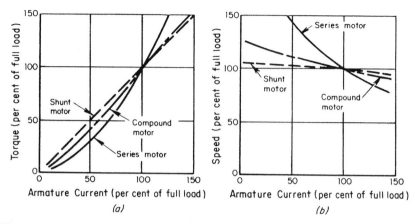

Figure 10-22. The compound-wound motor (see Figure 10-16) results in compromise operating characteristics between the shunt and series types, depending on the relative strength of the two fields. (Courtesy of *Machine Design,* Motor Reference Issue.)

However, when various types of braking are required to bring it to a stop, those mechanical or electromechanical brakes described in Chapter 15 are applicable. In addition, the dc motor is capable of being slowed down by either *dynamic* or *regenerative* braking.

Dynamic Braking. When a shunt motor is driving a load, if the armature circuit is opened, a voltage can be read across its terminals. A moment before, it was the counterelectromotive force, now it is the usual output of a dc generator. But the armature has no driving power except for the energy stored in its rotating armature and driven load. A resistance connected across the armature causes I^2R loss which quickly absorbs the energy and slows down the motor. At low speeds the generated voltage is so low, as is the I^2R loss, that the deceleration effect becomes almost negligible. It is the friction of the motor and load that brings it to a final stop (see Figure 10-23).

Regenerative Braking. This involves causing the running dc motor to become a generator pumping power back into the supply line; for instance, if a motor and load are running at a certain speed and the shunt field current is increased, the motor will naturally slow down, but if the WR^2 or momentum of the load is such as to try and maintain its speed, momentarily its counterelectromotive force is higher than the applied voltage and power is pumped back. This decelerates the motor more rapidly to its new reduced speed.

This is only a simple illustrative case of what can be accomplished

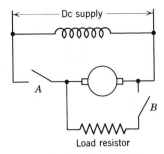

Figure 10-23. Connections for dynamic braking. Under normal running conditions switch A is closed and B is open. When A is open and B is closed, the armature, acting as a generator loaded on the resistor, continues to rotate. The energy stored in the rotating armature is dissipated as heat, thus reducing the armature speed rapidly. The switching is usually accomplished by push buttons actuating magnetic contactors.

by more complex control equipment, but because its action depends on pumping power back into the supply line, its application is limited by the capacity and type of supply.[1]

10-15   Control: Standardization

Since the dc motor is usually applied because of its flexibility of speed, it is nearly always used with a variety of control equipment. This is a field of technology in itself, which is thoroughly standardized by manufacturing groups, but will be neglected here with one exception.

Large-horsepower ac motors can usually be connected directly to the line by simple manual or magnetic switches. (They may be provided with overload or undervoltage protection.) The motor size and its starting current allowed may be fixed by local conditions.

But a dc motor beyond fractional-horsepower sizes must usually be "eased onto" the line. Not only may its large starting current have harmful effects on the relatively low capacity of its supply but the large armature current can cause vicious sparking and burning at the brushes with the possibility of pitting the commutator segments.

---

[1] The most readily understood case of regenerative braking would involve the now obsolete interurban and trolley cars (600 v dc) which, on going downhill, could have the field strength adjusted so as to pump power back into the supply system and maintain the desired safe speed. Perhaps this will become a useful factor in the electric automobiles planned for the future.

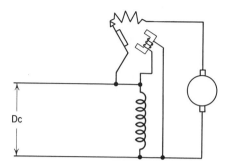

Figure 10-24. An elementary "starting box" for a dc motor offering undervoltage protection.

Starting equipment in its simplest form is a manual switch for introducing resistance in steps between the armature and its supply (see Figure 10-24). The lever on the resistance is usually turned against a spring and finally held in running position by an electromagnet. If the coil of this magnet is connected across the line, a power outage releases the lever to the off position. The reactivating of the line does not then start the motor without benefit of protection.

It might also be pointed out that dc motors from fractional horsepower up to 200 hp are built in frames with standardized mounting dimensions so that they interchange with similar ac motors (see Chapter 15).

10-16   Sources of Direct Current: Large DC Motors

A generation ago any self-respecting manufacturing plant that performed a variety of machine operations was equipped with an ac motor-driven dc generator. Naturally the capacities varied greatly, depending upon the need. Direct-current power was distributed to that portion of plant where variable-speed control was necessary in the manufacturing processes. Many of these generators are still in use, mainly because the equipment is still in operating condition. Others have been replaced. But bear in mind that we are thinking of an average plant. Steel mills and other processing plants still use dc motors of hundreds and thousands of horsepower. The electrochemical industries are examples where great dc capacity is required.

Two types of rotating equipment are available for such supply. The motor-generator set, as mentioned, offers an ease of control as compared with its competitor, the synchronous converter. The latter takes up less space and is usually less expensive per kilowatt. It will be described only briefly.

Consider a dc armature wound as usual and equipped with the conventional commutator. Suppose that this winding were tapped at suitable positions and three leads brought out to slip rings. The winding could then be equivalent to a three-phase delta-connected motor or alternator, looking at it from that end.

Place this armature in a stationary dc field, and with three-phase power applied to the slip rings, it can run as a synchronous motor. The current through the windings is "picked off" by the commutator and brushes and, thereby, rectified so that direct current is available at the brushes. The winding carries both ac input and dc output components of current. The schematic arrangement is shown in Figure 10-25, which indicates that it can be a somewhat versatile machine. It is chiefly a source of direct current at large outputs.

10-17 Metallic Plate Rectifiers

Copper Oxide. It was discovered some years ago that a copper oxide coating on a disk of copper will permit current to flow from the oxide

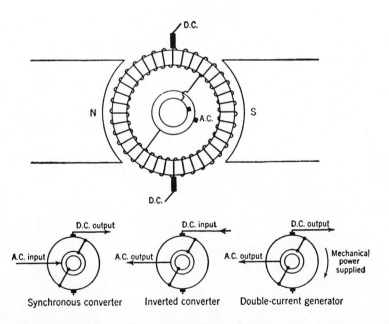

Figure 10-25. Schematic representation of a synchronous converter as a source of direct current and how it can be used for other purposes.

to the disk, but not in the opposite direction (see Figure 10-26). A perfect rectifier is one which has zero resistance in one direction and infinite resistance to current flow in the opposite direction. It will be noted that the copper oxide film is far from perfect, because 10 v, for instance, will cause either 2 or 10 ma of reverse current to flow, depending on the temperature. However, this is much less than the opposite flow.

For operation on higher voltages, stacks of these units are built up with cooling fins inbetween.

Selenium. These rectifiers act in a similar manner. Here a selenium coating is applied to one side of an iron or aluminum disk and a spring washer is bolted to the selenium side. Note the characteristics in Figure 10-27. This rectifier is capable of handling larger currents at higher voltages and is less susceptable to temperature change. Series-parallel combinations of these cells have been built up for outputs of about 100 kw.

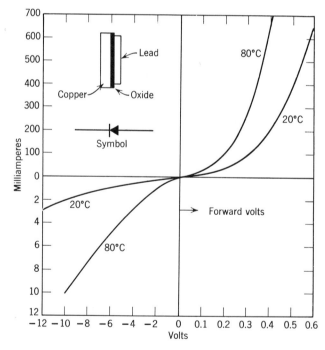

Figure 10-26. Characteristics of a copper-oxide rectifier with an active oxide film of 1 in.^{2}. Current flows in the direction of the arrow in the symbol.

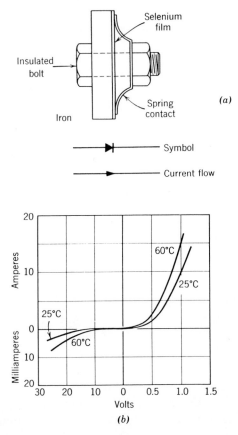

Figure 10-27. Voltage-current characteristics of the selenium type rectifier.

10-18   Connections

These metallic rectifiers are valves permitting current to flow in one direction and, for practical purposes, blocking a reverse current. It would appear then that, if connected in series with an ac line or through a transformer, we could obtain a dc output. This is exactly what happens, as indicated in Figure 10-28a, except that it results in a pulsating but unidirection current and is not too effective because only one-half wave of current appears. Four units connected in the "bridge" circuit as shown in Figure 10-28b give full-wave but pulsating current output.

Assume in Figure 10-28b that a shunt motor is connected across the output lines xy. With a usual field rheostat this would represent the simplest method of operating an adjustable-speed dc motor from an

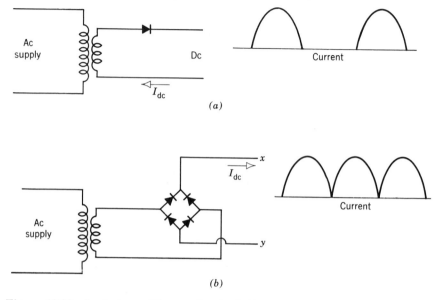

Figure 10-28. A single rectifier results in half-wave output as shown in (a). Circuit (b) represents a "bridge" arrangement with full-wave rectification. Volts available for a resistance-type load and resulting currents are shown.

ac source. Depending on the starting current, a protective resistance will have to be used in the armature circuit. The inductance of the windings help to smooth out the input to a limited extent.

This is the simplest case of dc motor operation from an ac supply. We shall see how these operations become more sophisticated.

10-19   Electronic Tubes and Elementary Circuits

Electronic tubes and solid-state electronics are so tied in as both the source of direct current for motor supply and the complex control of their operating functions that a brief description of these devices is included.

In the early 1880's Edison observed that, if a metallic plate was placed in an evacuated tube containing a heated lamp filament, there was a measurable flow of electrons from the filament to the plate. This has been called the Edison effect[2] (see Figure 10-29).

---

[2] It has been said that Edison invented dozens of useful devices but discovered only this one fundamental principle in electricity. Had further work been done with it, the field of electronics might have been advanced by 30 or 40 years.

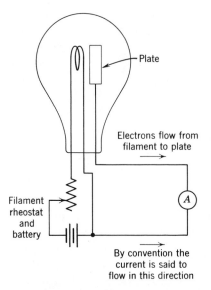

Plate

Electrons flow from
filament to plate
—→

Filament
rheostat
and
battery

A

—→
By convention the
current is said to
flow in this direction

Figure 10-29. Circuit illustrating the Edison effect. Electrons flow from the heated filament to the plate. Increase in filament current and temperature increases the plate voltage and current but not in a straight-line relationship. Basically this is a diode (two element) rectifier.

It may or may not be a new conception to the reader to learn that current flow as pictured in usual power circuits is just backward from the actual flow direction. This all started with Franklin with his concept of positive and negative changes, with the flow from positive to negative. But later it was discovered that current is a flow of negative electrons, which was movement in the opposite direction from the early conception. This can be ignored, except in dealing with flow in vacua, gases, or through electrolytes. By the ordinary concept, while current flows in the diode from plate to filament, negative electrons are actually moving in the opposite direction.

The Diode with an AC Supply. Figure 10-30*a* shows a common method of producing half-wave rectification with a diode. In Figure 10-30*b* full-wave rectification results with the use of a mid-tapped transformer secondary. This latter function can also be obtained by using the bridge circuit of Figure 10-28*b*, employing four diodes.

The Triode. This is a tube constructed essentially like the diode, except that a third screenlike element is added between the filament and the

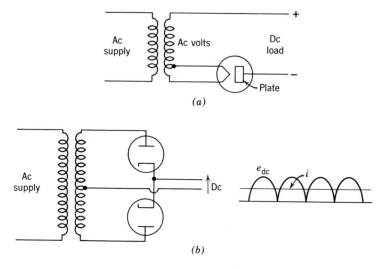

Figure 10-30. Half- and full-wave rectification by diodes. In (*a*) the symbolic representation of the diode follows the practice of some years ago. The modern symbols for the diode is shown in (*b*). The hooklike filament must have its other terminal connected to a filament supply. (*a*) Half-wave rectification. (*b*) Full-wave rectification, showing also how the current is smoothed on a load consisting of resistance and inductance.

plate. Electrons emitted from the filament must pass through this screen or grid.

An elementary triode circuit, using a completely dc supply and control, is shown in Figure 10-31.

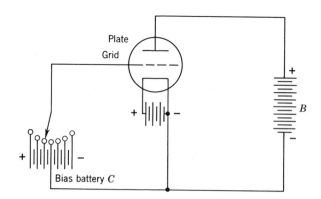

Figure 10-31. The triode showing how the grid can interrupt or amplify the electron flow to the plate, depending on the negative or positive voltage from the bias battery (loads have been omitted).

10-20   The Thyratron[3]

This is a triode-type tube filled with mercury vapor (or other gas) and widely used as a rectifier for a variety of capacities including large power circuits. The grid circuit is used only to start the plate circuit. The plate current can be interrupted by opening the plate circuit or reducing the plate voltage. A voltage drop occurs in the arc, which is relatively more important in lower voltage circuits.

Refer to Figure 10-32. Here, again, a battery is used to adjust the grid bias. But note in Figure 10-32a the effect of the voltage drop on the load voltage and the critical voltage E_c, which at any point between x and y is great enough to start or maintain the plate current. E_g can be so adjusted that the firing time can occur only between x and z. It is this quality that we wish to explore further.

Phase Shifting. Figure 10-33 shows a circuit in which the grid potential and its *position* are determined by the transformer secondary-voltage equation.

It is useful to note that the thyratron is controlled by keeping the grid too negative to permit the plate to conduct until the desired firing point. Thus anytime that the cathode-to-grid voltage becomes less negative than the critical value for a period sufficiently long for the gas to ionize, conduction takes place if the anode is positive with respect to the cathode.

With this general concept, any circuit that produces a cathode-to-grid voltage going positive after the cathode-to-anode voltage goes positive

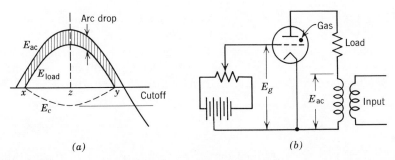

(a)   (b)

Figure 10-32. The thyratron triode circuit showing the interval voltage drop and the cutoff voltage E_c.

[3] Portions of this material are from Puchstein, Lloyd, and Conrad, *Alternating Current Machines,* John Wiley & Sons, 1954, p. 593, 3d ed.

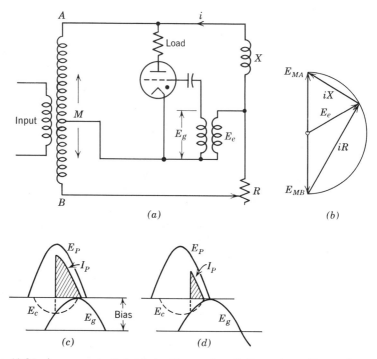

Figure 10-33. An ac controlled triode. By varying R the voltage E_e can be made to vary approximately 180°. This in turn changes the phase position of E_g with respect to the plate voltage and cuts off varying portions of the load current. This control is common, although a variety of phase-shifting circuits is possible.

can control the thyratron. If this delay is adjustable, it may be thought of as a phase-shift circuit or control circuit.

The waveshapes used vary, but include sine waves, exponential pulses, peaking transformers, and dc batteries. The battery is most marginal and functions on the principle that the critical grid-firing voltage becomes more negative with time during the first 90° of a sinusoidal cathode-to-plate voltage. Control is limited to the first 70° or so.

10-21   Semiconductors

This is a type of electronic device somewhat related to the metallic rectifiers but capable of all of the control functions of the vacua or gas-filled diodes and triodes. The theory involves the crystalline structure of germanium and silicon, either of which are used in this general class of devices known as *transistors* or *solid-state diodes*.

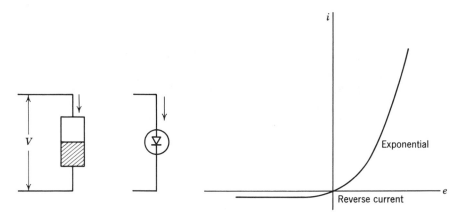

Figure 10-34. A semiconductor junction diode and its symbolic representation.

Either of these crystals is formed and treated in a manner which includes certain additives, giving in one case an excess of negative electrons which then becomes an n-type crystal. A different treatment for the same basic crystal leaves it full of "holes" (speaking molecularly in the vernacular of the electronics engineer) and it becomes a p-type crystal. A junction of the two form a rectifier which has a "forward and backward" current characteristic, somewhat similar to those of Figures 10-26 and 10-27 (see Figure 10-34).

Whether used as a simple rectifier or as a more complex control device, they are much smaller than the tubes which they replace and need no power for a heated filament. Maximum voltage before breakdown is a serious handicap, but their current-carrying capacities are large. However, germanium cannot be operated above a junction temperature of 100°C, whereas silicon is limited to 200°C. Heat-radiating disks and fins, called *heat sinks,* are used to help maintain operative temperatures. In the amplifier the transistor is basically current controlled, but otherwise similar to the vacuum triode or multigrid tube. The characteristics of a transistor are shown in Figure 10-35.

10-22   Silicon-Controlled Rectifiers

This is a form of semiconductor which might be described as a multiple-junction p-n-p-n buildup, which has all of the qualities of the triode tubes such as the thyratron. Its multiple uses have resulted in it being built in a variety of forms in large enough production so that it is

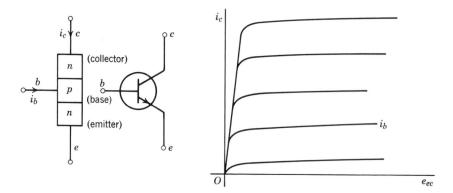

Figure 10-35. Output characteristics of an n-p-n transistor.

comparatively economical. Of its many uses, we shall show only how it substitutes for the thyratron.

Similar to the thyratron, the SCR is an open switch to positive cathode-to-anode voltage until a signal at the gate turns it on. It then operates independently of the gate or grid, and is stopped only by reducing the anode-to-cathode current to zero for a short period of a few microseconds while control is regained.

The SCR in keeping with its solid-state nature and transistor background is sensitive to current signals at the gate whereas the thyratron was voltage sensitive. Thus some power must be expended to trigger the SCR, and the cathode-to-gate voltage must be positive for the "turning on" period of a few microseconds.

A special problem exists for the SCR. Since power is supplied to the gate at firing, it must not be too great or for too long a period such that the resulting heat cannot be dissipated.

The device is particularly sensitive to having positive gate current while the anode-to-cathode current is negative. Internal operation is most simply modeled by considering the p-n-p-n to be two transistors in parallel (see Figure 10-36).

Positive cathode-to-anode voltage, e, appears primarily across the n-p junction from b_1 to c_1 and/or c_2 to b_2, as two normally operating transistors. Any current at g represents positive base current to b_2 and supports a larger collector current in c_2. However this current is a positive base current for b_1 and supports a larger collector current in c_1. This current, in turn, is more positive base current for b_2, and the process is self-supporting if the two gain factors $B_1 = i_{c_1}/i_{b_2}$, $B_2 = l_{c_2}/i_{b_2}$. The value of B's depends on current level, in general increasing with i. Thus a pulse at g can start a process then self-sustaining.

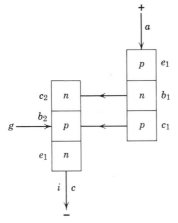

Figure 10-36. Showing two transistors in parallel, as a means of representation.

Control is accomplished in any manner that can supply the current pulse at i_g properly timed with respect to e. A few firing circuits are shown in Figure 10-37. More complex circuits may be used, but the same principle is utilized.

The trigger diode and the unijunction (UJ) mentioned in Figure 10-37

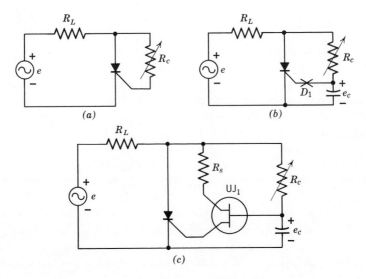

Figure 10-37. SCR firing circuits. (a) Fires when $i_{g(critical)} = e/R_i + R_c$. (b) Fires when e_c is large enough to break down diode D_1. (c) Fires when e_c changes to critical value of gate on unijunction UJ.

both exhibit the nonlinear property of voltage breakdown from a voltage v_2 to a voltage v_1, with nearly short-circuit characteristics between. The diode voltage values are determined in original fabrication, whereas the breakdown value for the UJ is adjustable in terms of the voltage applied across it.

In each case the device serves to discharge a portion of the stored energy in a capacitor, and the resulting current pulse triggers the SCR. In many applications the components are not easily divided into the blocks of Figure 10-37 but have overlapping functions and effects. Economy usually dictates that simplicity and compactness dominate over perfect performance.

Most practical circuits must include also various protective devices:

1. Starting current limit.
2. Fail safe.
3. Time delay if any heaters are involved.
4. Voltage-surge protection.
5. Voltage and current protection for the electronic parts.
6. Ventilation and cooling.

10-23 DC Motor Drives: A Summary

Machine tools from a simple lathe to complex automated devices have been manufactured for some years with dc drives because of their comparative ease of speed control.

Some machinery is sold with each unit having its own motor-generator (MG) set. Thus one has an ac motor, a dc generator, and a controlled dc motor for each machine tool. (Historically this was known as the Ward Leonard system.) Control is obtained not only by changes in the dc motor field and armature circuit but also by the dc generator field and voltage control. Many of these are still being built.

Banks of metallic rectifiers of the copper oxide and selenium type have been employed for motor drives, although their major use has probably been for battery charging and small elements of the electrochemical industry.

During and after World War II the electronic-tube circuits became a source of direct current for machine tools and other factory equipment. (Actually, the mercury arc rectifier had been in use for a considerably longer period than that as a source of direct current for the now defunct interurban and city street railway systems.)

Many of these machine tools are still in use. Their sale was initially handicapped by the fact that the average plant electrician, well trained in motor and control maintenance, was confused by finding his power supply cabinet full of small resistors, capacitors, and glowing tubes explained by a circuit diagram larger than his desk. His usual reaction was negative. A program of training was required.

Most recently we have used the semiconductors and especially the SCR's with their steadily reducing size and costs. Here, again, special training is necessary for maintenance and repair. As a source of rectified alternating current and as a quick response control, they offer what seems to be an infinite variety of possibilities for dc drives.

Most electronic control of dc motors is accomplished by providing a variable-magnitude dc component of voltage by controlled rectification of a sinusoidal ac supply. Two functions are carried out simultaneously. First, the circuit components including the SCR's or thyratrons are so arranged as to provide only unidirectional flow of current to the motor from the alternating supply. Various common circuits such as half-wave, full-wave center-tapped, single-phase or three-phase bridge, and others are used.

Secondly, the use of controlled rectifiers, that is, SCR's, thyratrons, or ignitrons, makes it possible to control the point in the ac cycle at which the device fires. (However, control can only be exercised to keep the device from firing during the initial part of the period in which it would have begun conduction as a diode.) Thus the maximum dc voltage is equal to that provided by the basic diode circuit and may be reduced from this maximum to zero.

The resulting control is quite like that obtained with a variable dc supply, but it has an added component of harmonics which may vary widely in magnitude and frequency as a function of the circuit chosen.

Various alternatives exist but are not used so frequently:

1. Two control devices back to back, providing a variable alternating supply that consists of controlled parts of the regular supply voltage. Both amplitude and phase vary, and appreciable harmonic content is introduced.

2. Electronic devices may be used to substitute for the conventional commutator as well as for the function of rectification.

3. Variable-frequency ac voltage may be provided by means of an electronic inverter, thus providing for variable-speed ac motors. A few examples of basic speed-control circuits are described in the following.

Consider a dc motor required to maintain nearly constant but adjustable speed over wide ranges of load. Such performance usually requires the addition of another basic concept, namely, feedback. This function

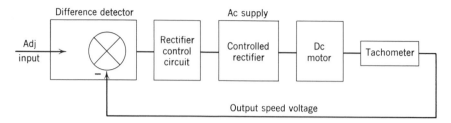

Figure 10-38. Components of a controlled dc motor.

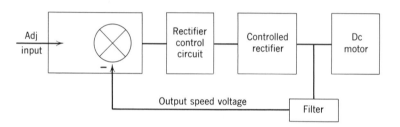

Figure 10-39. Components of a controlled dc motor by back emf.

is sometimes referred to as *regulation* or *speed control*. In each case, some knowledge of the actual speed is compared with some form of reference or desired value, and an error, if existing, is used to reduce the difference. This is specifically negative feedback.

In the dc motor application the measure of actual speed is usually obtained by one of two means—a tachometer of some form providing a voltage proportional to speed or an attempt is made to use the back emf of the motor as an indication of speed, other factors being held constant. The reference may be actual, but is more often a virtual condition or reference voltage. A block diagram of parts is shown in Figure 10-38.

If the tachometer is omitted in favor of the back emf, the diagram differs only in that the output speed-voltage signal is obtained by filtering from the input motor terminals (see Figure 10-39).

Chapter 11

Universal Motors

11-1   Definition

A universal motor is one which is capable of being run on either direct or alternating current with approximately the same operating characteristics. In the latter case it is assumed that the frequency is 60 cycles or below.

11-2   The AC Shunt Motor

We found in dealing with dc motors that reversing the motor leads with respect to the supply lines did not reverse the motor. It was necessary to reverse the armature or field leads with respect to each other to cause a reversal in motor direction.

Consider now a *shunt motor* connected to an ac line. As the ac cycle goes to a negative value, both armature and field connections have the same reversal of current and the motor operates in the same direction.

But there are several things wrong with such a motor. The field with its many turns of wire linking the main flux system has a high inductance. Its current will lag the voltage by a large angle. The inductance of the armature is relatively less. The result is that, when the field current and flux are a maximum, the armature current is on a lower part of its cycle. When the armature current is a maximum, the field flux is relatively small. This results in little torque.

However, if a capacitor is connected in series with the shunt field, the current in that circuit can be brought almost into phase with that of the armature. Also the effective voltage on the field winding is raised. The result is a motor which operates with comparatively constant speed.

The only value of this type of motor is that it can be made to run at higher speed than conventional induction motors. Such designs are not generally available.[1] It cannot qualify as a universal motor.

11-3   The Series AC Motor

The series motor can be built as a truely universal type. When connected to an ac supply, the current in both field and armature are naturally in phase, and the reaction between the field flux and the armature current in producing torque is practically the same as though dc were applied.

A few differences appear. Some small dc motors can be built with solid iron or steel as the yoke and motor shell, and even with solid field poles. The series universal motor must always be built of laminated steel because of the eddy currents which would otherwise cause excessive losses on alternating current.

Commutation is a little worse on alternating current. This is because the coil short-circuited by the brush during rotation is not "dead" but has a voltage induced in it by transformer action. The resulting sparking, although still not excessive, does shorten the brush life.

When running on direct current the useful voltage applied to the armature and field is the line voltage minus the IR drops. But when alternating current is applied, the natural reactance of both the armature and field gives the motor an additional IX drop. Added at right angles to the IR drop (as in any such ac circuit), the resulting impedance drop leaves less useful voltage. This reduction in voltage in both armature and field, as compared to direct current, means that the motor operates at a slower speed. Comparative speed-torque curves are shown in Figure 11-1.

11-4   Brush Shift and Commutation

As in the case of the dc motor, the field built up by current in the armature is at right angles to that of the field poles. The resultant flux is distorted, shifted backward against the rotation of the motor. For improved commutation it is again necessary to give the brushes

---

[1] The author is aware of a few special cases where these motors filled a need and were built in small quantities. Although it is *not* a universal motor, it is described here as an answer to the question, often brought up, concerning its characteristics.

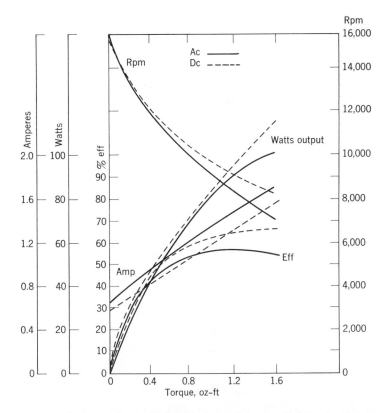

Figure 11-1. Typical performance curves of a $\frac{1}{12}$ hp 115 v nominal 10,000 rpm universal motor. The full-load torque is 0.7 oz-ft. Starting torque is 5.1 oz-ft on ac and 18.3 oz-ft on dc. Starting amperes are 3.7 ac and 8.4 on dc.

a backward shift. For reasons of symmetrical construction of the motor body and/or the motor end heads, the brushes are usually located symmetrical with respect to the poles and the armature winding is displaced as an equivalent to brush shift (see Figure 10-12).

Commutation can result in severe brush wear and shortened brush life because of the increased sparking with alternating current. Also the high speed of the commutator adds to the wear and the reduced life. Another factor found in universal motors is a tendency for brushes to "bounce" and chatter, yet increased spring pressure to reduce this may result in faster wear.

In general, the brushes used on any universal motor design have been life tested and the correct grade has been chosen for satisfactory operating characteristics. Replacing worn brushes by anything that apparently

Table 11-1 Typical Ratings and Sizes of Universal Motors

Horsepower	Rpm	Shell Diameter
$\frac{1}{50}$	5,000	3
$\frac{1}{20}$	10,000	3
$\frac{1}{20}$	5,000	$3\frac{3}{8}$
$\frac{1}{12}$	10,000	$3\frac{3}{8}$
$\frac{1}{12}$	5,000	$3\frac{3}{8}$
$\frac{1}{8}$	10,000	$3\frac{3}{8}$
$\frac{1}{8}$	5,000	$3\frac{7}{8}$
$\frac{1}{6}$	10,000	$3\frac{3}{8}$
$\frac{1}{6}$	5,000	$3\frac{7}{8}$
$\frac{1}{4}$	10,000	$3\frac{7}{8}$
$\frac{1}{4}$	5,000	$3\frac{7}{8}$
$\frac{1}{3}$	10,000	$3\frac{7}{8}$
$\frac{1}{3}$	5,000	$3\frac{7}{8}$
$\frac{1}{2}$	10,000	$3\frac{7}{8}$
$\frac{1}{2}$	5,000	$5\frac{7}{16}$
$\frac{3}{4}$	7,500	$5\frac{7}{16}$
1	10,000	$5\frac{7}{16}$

Figure 11-2. A typical heavy-duty universal motor with resilient base. (Courtesy of Robbins & Myers, Inc.)

fits from a service stock can result in short life, excessive commutator wear, sparking, and unusual noise.

Brush life may be anywhere from 100 to possibly 800 hr. Although these periods seem short, we shall see that for the equipment on which universal motors are most commonly used this may represent from 5 to 10 years of useful life.

11-5 Complete Motors and Motor Parts

Complete motors represent a relatively small part of the business. *Motor parts* are sold in large volume.

Motor sizes and ratings are not standardized aside from a few military specifications. A typical lineup of ratings and sizes that might be found in a motor manufacturer's stock are given in the Table 11-1.

Standard rotation is counterclockwise facing the end opposite the shaft extension. Motors must be especially built for opposite rotation.

Figure 11-3. Cutaway view of a set of universal motor parts. Note the relatively small cross sections of parts of the magnetic circuit. (Courtesy of Robbins & Myers, Inc.)

Figure 11-4. These motor parts with a field diameter of 3.687-in. drive this medium-sized drill.

A typical construction is shown in Figure 11-2. Here a cradle or resilient base mounting is shown, with synthetic rubber rings around the hub mounting. This is to reduce vibration from being transmitted to the machine in which the motor is mounted. Rigid base or face mountings are also common.

A cutaway view of a set of motor parts is shown in Figure 11-3. Only one of the two leads are shown with a spring connector that fits around the metal part of the brush holder. These units with fan and a set of cartridge brusholders represent a typical set of parts which might be sold to the manufacturer of portable electric tools.

Examine the drill and motor parts of Figure 11-4. It can readily be seen that for the manufacturer of any portable tools it is necessary to know all of the important dimensions of the motor parts available before he can design the die-cast aluminum (or plastic) components which make up his unit. Therefore these important dimensions are a National Standard (see Figure 11-5). Of the dimensions tabulated we shall ignore all except the possible diameters BH. They are 2.125, 2.437,

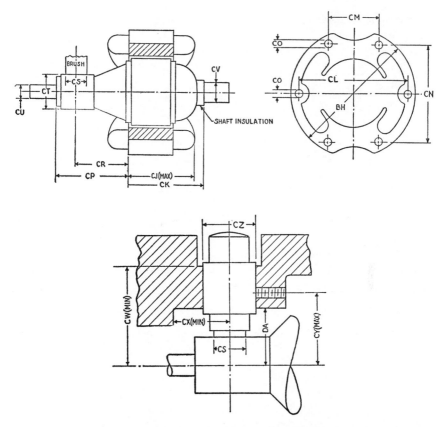

Figure 11-5. These dimensions are standardized by NEMA for universal motor parts.

2.875, 3.187, 3.687, and 4.375 in. Tolerances are +0.000 and −0.002 in. These standards were developed by NEMA in 1948. The universal motor parts business had been in existence for many years before that and, as a result, other diameters, developed earlier, are still available.

11-6   Applications

Devices using universal motor parts comprise a long list.

Vacuum cleaners	Blenders
Floor scrubbers and polishers	Coffee grinders (domestic)
Sewing machines	Electric drills
Food mixers	Hedge trimmers

Electric lawn mowers (some types)   Saws: Disk
Routers   Saber
Sanders: Chain (up to 3 hp)
    Disk
    Belt

If we stop to consider where these devices are used, it will be realized that the universal characteristics of these motors is of almost no importance. (Who has direct current in his home?)

These universal motors are chosen because they offer the most horsepower per pound owing to their high speed. It was pointed out in discussing induction motors that the slower the speed the larger the motor. Naturally the converse of this is true. As horsepower depends on speed and torque, a speed of, say, 12,000 rpm needs only a torque of 0.438 lb ft for 1 hp. Torque, in turn, is proportional to flux per pole. With reasonable flux densities the pole flux can and must be comparatively small in these motors of only a few inches in diameter.[2]

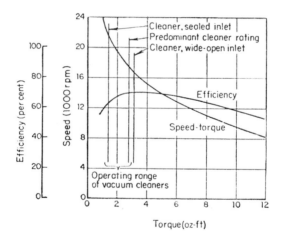

Figure 11-6. Speed and efficiency versus torque for $\frac{9}{16}$-hp 17,500-rpm universal motor used in a domestic vacuum cleaner. (Courtesy of *Machine Design,* Motor Reference Issue)

---

[2] Many portable *industrial* tools used almost constantly are powered by high-cycle three-phase motors parts. Frequencies of 120, 180, 360, and 400 are common. Such motors do not require the brush and commutator service which would result otherwise. But they, naturally, require a special power source.

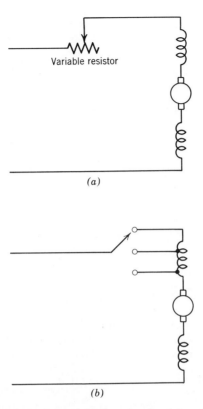

Figure 11-7. Speed control of a universal motor by (*a*) series resistance and (*b*) tapped field.

The drooping speed characteristic of the series motor is also useful. On a drill, for instance, a small diameter drill bit operates at high speed. A large diameter bit should run slower and does so because it loads down the motor.

This characteristic is also useful in a vacuum cleaner. Its blower performance is such that when the inlet is about sealed off, it unloads and operates at high speed; with a wide open inlet it loads up and slows down (see Figure 11-6).

High no-load or run-away speed is not too serious in these applications. It will be noted that nearly all of these motors are connected to their load either directly or through gears. The gear and the ventilating fan help to keep the speed to a safe value. This is especially necessary on portable disk saws where safety rules limit the idle speed.

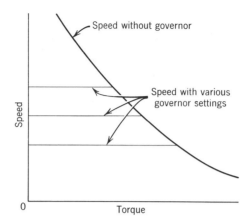

Figure 11-8. Universal motor with centrifugal governor. Note that at heavier loads the governor cannot maintain the speed.

11-7   Speed Control

The simplest method of controlling the speed of the universal motor is by a resistance in series with the motor (see Figure 11-7a). Also tapped windings can be used. They present the disadvantage of reducing the torque at low speeds.

On the other hand, silicon-controlled rectifiers as described in the preceding chapter can be designed into circuits to maintain better torque at low speeds.

Speed governors are available for these motors. They are essentially centrifugally operated switches which open and connect a resistance in the supply line when a predetermined speed is reached. This slows down the motor closing the contacts which short-circuit the resistance. The cycle keeps repeating itself rapidly so that the final speed is determined by the ratio of the times in which the contacts are open or closed. The controlled speeds are adjustable (see Figure 11-8).

Chapter 12

Mechanical Variations in Motor Construction, Bearings

12-1 Special Mechanical Constructions

A typical electric motor is one that is rigidly mounted on a base, has open ventilation, and is perhaps $1\frac{1}{4}$ to $1\frac{1}{2}$ times as long as its diameter. Actually millions of motors or motor parts are sold which are not like the foregoing description. We have already noted in Chapter 11 that universal motor parts represent a large segment of the motor business, sold merely as parts to be fitted into frames manufactured by suppliers of portable tools, vacuum cleaners, food mixers, etc.

Figure 12-1 represents one extreme in comparative dimensions. These are three-phase 180-cycle motor parts which drive a small unbalanced flywheel. Mounted in a long flexible tube, with no ventilation, they are the driving power for concrete vibrators. Stuck into a mound of freshly dumped concrete, the vibrations level it in a very short time.

Figure 12-2 shows the motor parts of a variety of designs known as *pancake* motors. These particular parts are also for three-phase operation and are used as elevator door openers, where there is little space between the elevator shaft and cage.

Another common variant involves motors in which the shaft and rotor are not in the center of the motor. Stator laminations and the body are flat on one side. When used with a circular saw, this enables a deeper cut to be made with any given diameter of saw blade.

The point is that motors are available in a variety of shapes and sizes, if the purchaser is willing to pay for all of the special tooling involved or if he presents a convincingly high enough market so that the motor manufacturer will invest in such tooling.

Figure 12-1. Three-phase 180-cycle motor parts, with a synchronous speed of 10,000 rpm. They drive an unbalanced small diameter flywheel as a vibrator. (Courtesy of Robbins & Myers, Inc.)

12-2   Environment Affects Motor Construction

The ordinary motor is expected to operate under the following conditions that might be classed as normal.

The temperature of the surrounding air should not be greater than 40°C (104°F). This is called the *ambient temperature*.

The atmosphere should not contain excessive dust, moisture, or fumes

Figure 12-2. "Pancake" type motor.

that will seriously interfere with the operation of the motor. (This, of course, is a very broad statement, subject to considerable personal opinion.)

The frequency shall not vary more than 5% above or below the nameplate value. This is rarely a problem.

The voltage shall not vary more than 10% above or below the nameplate rating. (This is a fairly frequent problem.)

The location of the motor shall not be more than 3,300 ft above sea level. This is mentioned because in the thinner air of high altitudes less heat is carried off by the ventilating system. Theoretically, motors should be designed for lower operating temperatures at normal elevations so that they will have some leeway if expected to be used at high elevations. This is recognized in large equipment, but the manufacturer of an electric dishwasher or an electric typewriter, for instance, does not demand a special motor when his product is sold in Denver (elev. 5200 ft).

Considering all of the foregoing factors, it can readily be seen that the motor used in the home or in the office is most likely to be operating under relatively ideal conditions. (We will see in Chapters 14 and 18 how special problems can sometimes modify these advantages.)

12-3   Special Environments

A motor construction so common that it can hardly be classed as special is one in which all of the body and head openings are in the lower half of the motor. This is called a *dripproof machine,* because it is assumed that any liquids or solids falling from an angle of not greater than 15° from the vertical cannot enter the motor. An example of this construction has been shown in Figure 7-17. Another example is pictured in Figure 12-3.

In the design of a line of motors, from fractional-horsepower up through large integral-horsepower sizes, most manufacturers construct the motor frames and end heads to meet this requirement and sell it as a standard or "off the shelf" construction.

A more severe restriction on the motor openings is presented by the *splashproof* requirements. For this motor the openings are so constructed that "drops of liquid or solid particles falling on the machine or coming toward the machine in a straight line at any angle not greater than 100° from the vertical cannot enter the machine" so as to interfere with its successful operation.

An example of this construction is shown in Figure 12-4. This is not only splashproof but also meets the requirements of the two following definitions:[1]

Guarded machine: An open machine in which all ventilating openings are limited to specified size and shape to prevent accidental contact with rotating parts (except smooth shafts).

Weather-protected machine: A Type 1 weather-protected machine in an open machine with its ventilating passages so constructed as to minimize the entrance of rain, snow and airborne particles to the electric parts and having its ventilating openings so constructed as to prevent the passage of a cylindrical rod ¾ inch in diameter.

In some motors it is not only important to restrict the size of the openings so that persons cannot become injured by pushing their fingers into the holes but also to keep out rats and mice. This is especially

Figure 12-3. A dripproof motor with openings in only the lower half. The rabbet mounting on the head has standardized dimensions used for mounting a variety of equipment. In this case the motor was designed for a "close-coupled" pump. (Courtesy of Robbins & Myers, Inc.)

[1] Many of the definition in this chapter are taken wholly or in part from the NEMA Standards, of which more will be learned in Chapter 13.

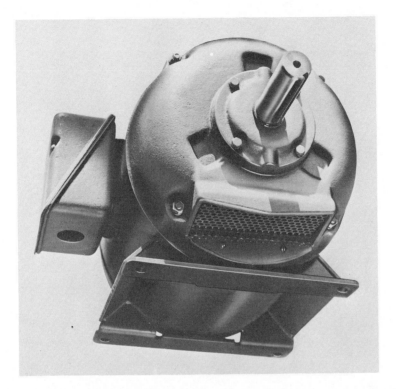

Figure 12-4. A splashproof protected motor. This mechanical design is intended for outdoor use. (Courtesy of Robbins & Myers, Inc.)

true in farm applications. These rodents have developed a taste for the insulation used on conductors. This seems to be especially true for some of the newer plastic insulating materials.

12-4   Effect of Ventilating Methods on Motor Construction

An *externally ventilated machine* is one in which ventilating air is forced through it by means of an external fan or blower. This is quite common on extremely large motors or generators but may also be used under special conditions. A good example is pictured in Figure 12-5. Here is a large, single-phase motor in which the size and weight were restricted by the user. Operating on 60 cycles, 230 v, it was cool enough at rated horsepower to satisfy the load requirements. But the motor and the device with which it was used were occasionally sold abroad

Figure 12-5. An externally ventilated motor. In this case the small motor and blower were added only to provide a cool-enough motor when operated on 50 cycles. (Courtesy of Robbins & Myers, Inc.)

where 50-cycle power is common. Operating on 50 cycles the motor overheated badly because of the slower speed of the built-in fan and because the flux density is increased in reducing the frequency, resulting in increased losses.

The small externally mounted blower and motor increased the volume of air through the motor, allowing it to operate cooler on 50-cycle supply.

A *pipe-ventilated machine* is so arranged that its inlet openings for the admission of air can be readily covered by a pipe or duct, leading air in from a distant position. This again applies most often to very large machines, but it can make a useful method of ventilation in other cases. We shall consider an example after discussing the next type of protection and cooling.

In Chapter 14 we shall see that motors are most effectively cooled by having a good blast of cool air forced through them, carrying the heat away from the windings and other components. But sometimes this air is so laden with fumes and/or abrasive or metallic dust that there is danger of corrosion or wearing away the insulation, thus eventually causing a breakdown between the windings. Or these fumes and/or dust may ruin the bearings.

The obvious thing to do is to *enclose* the motor, leaving no openings for ventilation. Such motors are not really airtight but their exposure has been greatly reduced. Unfortunately, cutting off the air causes excessive overheating, and the *enclosed motor* must be much larger in size to yield enough dissipating area so that the temperature rise will be acceptable.

Just as you can cool a spoonful of soup by blowing on it, so can you cool an enclosed motor by blowing air over as much of the outside surface as possible. Such motors are called *totally enclosed fan-cooled* (TEFC) and they are about the same size for a given horsepower and speed as an open motor; but they are expensive.

Figure 12-6 shows a ½-hp single-phase motor of such construction. In this cutaway view notice the cover or cowl at the left. On the back center of this cowl are a series of slots through which air is fed to the large fan. This air is directed by the cowl over the shell of the motor. This is sufficient to cool the motor even though no air passes through it internally. The fan blades on the rotor are to stir up the air inside the motor. (The "slot" in the bottom of the right-hand head is not an opening but is a groove to enable a grease fitting to be applied to the hub.)

An industrial-type TEFC motor is shown in Figure 12-7. Here fins are used on the heads to increase the heat-dissipating area. These are also built with cast iron or aluminum finned bodies for still better heat dissipation.

Encapsulated Motors. A comparatively recent addition to this class of protected motors is one in which the windings are encapsulated. This is defined as follows:

Encapsulated machine: An encapsulated alternating-current squirrel-cage machine is one having random windings filled with a insulating resin which also forms a protective coating. This type of machine is intended for exposure to more severe environmental conditions than usual varnish treatments can withstand. Other parts may require special protection against such environmental conditions.

The usual method of construction is to make molds which fit into

Figure 12-6. A small totally enclosed fan-cooled motor, cut away to show external fan and internal construction. (Courtesy of Robbins & Myers, Inc.)

the bore of the wound stator and over the end coils, allowing clearance between the coils and the inner surface of this latter section of the mold. A liquid plastic is forced into those openings so that all air spaces in the slots are filled and a uniform protective sheath is provided over the extended end coils (see Figure 12-8).

In many cases this protection is adequate for operating under conditions formely requiring TEFC motors. Encapsulated motors are considerably less expensive than the TEFC, but more costly than standard open motors.

12-5   Are TEFC Motors Oversold?

In the opinion of the author they are, in many cases. We hear statements by purchasing agents that for 30 years they have bought nothing but TEFC motors for their plant. Dust and fumes. This is reinforced by statements from their plant electricians that they have had a splendid record of operation using such motors. Of course, but how do they know that a modern, open motor would not now do as well with great savings in first cost? Great advances have been made in the quality of insulating varnishes in recent years. They are tougher, more abrasive, and moisture-resistant. Many of them are resistant to certain chemical action.

Encapsulated motors have already been mentioned as occasional substitutes for TEFC motors, but the pipe-ventilated machine can also be used in some instances. Figure 12-9 represents a motor in which clean outside air is drawn in by the motor fan through a duct as shown. A whole lineup of such motors was used in a plant in which abrasive and metallic dusts were present. The open motor and the duct work were considerable less expensive than TEFC motors.

Unless the problem is taken care of in the foregoing manner, the presence of airborne metallic dust still represents a good case for the

Figure 12-7. An industrial, totally enclosed fan-cooled motor. (Courtesy of Robbins & Myers, Inc.)

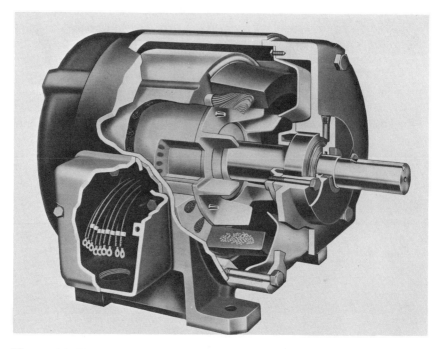

Figure 12-8. This encapsulated motor is dripproof and has a sealed insulation system that protects against detrimental atmospheric conditions. It can be used in some applications in place of a totally enclosed motor. (Courtesy of Allis-Chalmers.)

TEFC motor. As mentioned previously, layers of it building up on parts of the winding might lead to an electrical breakdown.

The Great Outdoors. Up to about 20 years ago, if a motor had to be used outdoors, a TEFC type was usually chosen. If an open motor was used, a shelter of some sort was built over it.

The oil-pumping industry probably pioneered in the use of open motors, otherwise unprotected, for outdoor service. Mechanically such motors are of standard construction, except that they are usually given special protective paints and antirust treatment, and have screened ventilating openings to keep out snakes and rodents.

Hundreds of thousands of such motors are in use in such places as Louisiana, west and north throughout the "dust bowl" and on up into Wyoming.

More recently, segments of the air-conditioning industry, using motor-driven fans and/or pumps for condenser cooling on top of office buildings, etc., are changing to "weatherized" open motors.

12-6   Ball Bearings

Sealed ball bearings and, in general, better bearing protection in motors have been of considerable importance in permitting successful applications where dust and fumes are prevalent.

Figure 12-9. This pipe-ventilated motor is a standard motor reassembled with one head upside down and a metal duct attached to the opening. The standard internal fan brings clean outside air in to ventilate the motor in an atmosphere that is laden with metallic abrasive dust. (Courtesy of Robbins & Myers, Inc.)

In general, ball bearings for motors fall into three classes:

1. *Open bearings.* The head castings and cover plates are machined with small clearance between the shaft and stationary parts. "Slingers" might also be used on the shaft, which by centrifugal force throw dirt-laden air away from these clearance openings. Such motors are usually supplied with grease fittings on top of the hubs and pressure-release fittings on the bottom of the hubs so that fresh grease can be added from time to time with a grease gun (see Figure 12-10).

2. *Shielded bearings.* Seals are provided on one or both sides of the bearing. These are usually metal and are affixed to the outer or stationary bearing race. A slight clearance is provided between the seals and the rotating race. These are usually quite adequate for keeping in the grease and keeping out dirt (see Figure 12-11a).

3. *Sealed bearings.* In these bearings there is actual running contact between the seal and the turning inner bearing race. Again they can be purchased with a seal on one or both sides (see Figure 12-11b).

As in the case of the shielded bearings, they are usually purchased ready packed with grease of a type specified by the user. The most common "special" is a grease developed for high-temperature operation.

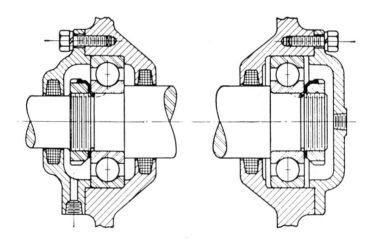

Figure 12-10. Open-type ball bearings used in the hubs of a motor. Illustrated here is a condition in which the shaft extension (*left*) is not permitted to move back and forth ("end play"). Note that that bearing is fixed with respect to the hub and shaft. Because shafts expand with heat, the other bearing (*right*) must be free to shift axially. The areas around the bearings are filled with grease. Note the seals on the shaft which keep grease in and dirt out. This is comparatively expensive construction. (Courtesy of Marlin-Rockwell Corp.)

(a)

(b)

Figure 12-11. Shielded and sealed ball bearings, all of standard sizes, are inter-changeable with open bearings. Packed wth grease when installed, they are most frequently used in motors with no provision for relubrication. (*a*) A shielded ball bearing with slight clearance between shields and the running inner race. (*b*) This sealed ball bearing maintains running contact between the seals and the inner race. This adds slightly to the bearing friction. Synthetic rubber with a steel core form the seal in this case. (Courtesy of Marlin-Rockwell Corp.)

12-7   Sleeve Bearings

These bearings vary in type and construction, depending largely on motor size.

Nearly all subfractional-horsepower motors use self-aligning porous bronze bearings containing more or less graphite in some cases. These might be found in motors up to $\frac{1}{8}$ or $\frac{1}{2}$ hp, but more commonly in ratings below that.

Essentially they are spherical in shape with a uniform inner bore. The oil permeates the bearing from a felt-lined oil reservoir. The outside surface usually has a groove which matches a projection in the motor hub to keep the bearing from turning. Its self-aligning feature, however, permits the bearing to be perpendicular to the motor head, which is, of course, desirable, or swung in any direction through an angle of, say, 5 to 10°.

Figure 12-12 presents a cross section through an entire aluminum die-cast head, showing assembly of all components. Note how the bearing rests in a semispherical recess (not machined) and is also held in place by a shaped washer, behind a cover or oil return cap on the inside of the motor. These two latter-formed metal parts are held in place by crimping over some of the die-cast hub.

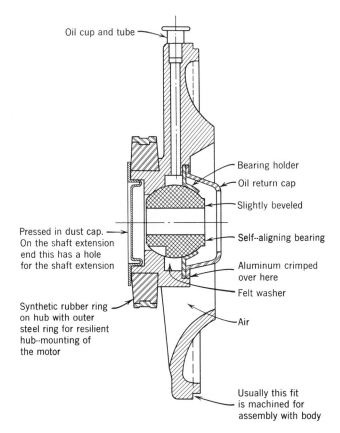

Oil cup and tube

Bearing holder

Oil return cap

Slightly beveled

Pressed in dust cap.
On the shaft extension
end this has a hole
for the shaft extension

Self--aligning bearing

Aluminum crimped
over here

Felt washer

Synthetic rubber ring
on hub with outer
steel ring for resilient
hub--mounting of
the motor

Air

Usually this fit
is machined for
assembly with body

Figure 12-12. The assembly of a self-aligning bearing in a die-cast aluminum end head, also showing other parts. (The writer would estimate that more than 100,-000,000 motors have been built with this type of bearing.)

Small, Rigid Bearings. These may be built of porous bronze, but frequently use steel-backed babbitt. The latter, of course, forms the bearing surface and usually has grooves spreading out from an opening in the cylindrical surface of the bearing. Oil-saturated felt or wool packing, in contact with the rest of the oil reservoir, presses through this bearing opening and, with the grooves, forms a means of distributing oil along the inner bearing surface (see Figure 12-13).

Closer machining tolerances are required in this construction, for if the heads are slightly "cocked," the bearings may bind on the shaft instead of aligning themselves.

Sleeve Bearings for Large Motors. In large integral-horsepower motors, bronze bearings, rigidly mounted, are used as previously described. But the top center area of such bearings is cut away so that a ring of comparatively larger diameter is free to ride on the shaft, with its bottom portion dipping into the oil reservoir. When running, this ring rotates rather slowly, owing to slippage, but continuously brings up a fresh supply of oil to the shaft, which distributes itself throughout the bearing clearance. Such a motor can be used in only one position.

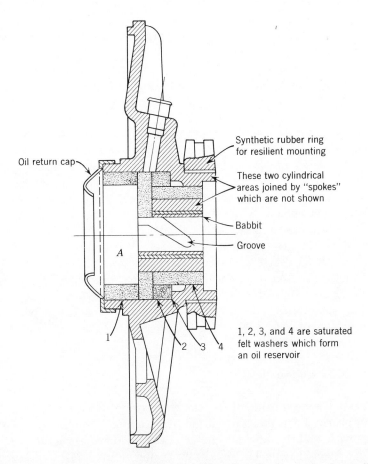

Synthetic rubber ring for resilient mounting

Oil return cap

These two cylindrical areas joined by "spokes" which are not shown

Babbit

Groove

A

1, 2, 3, and 4 are saturated felt washers which form an oil reservoir

1   2   3   4

Figure 12-13. Assembly of a steel-backed babbett bearing in an aluminum die-cast head. That portion of the shaft operating in area *A* is equipped with a washer or "oil slinger" to prevent oil from creeping along the shaft into the motor. It is returned to the oil reservoir through felt **1**. The recess in the right side has a cap with or without a seal inside.

12-8   Comparisons

It is slightly oversimplifying the problem to say that ball bearings are used wherever there is considerable side or end thrust (belt, chain, or gear drive), and sleeve bearings are used wherever noise or cost is a factor. Yet this is basically the case, with a number of exceptions.

Only occasionally we find ball bearings on a *subfractional-horsepower* motor, unless it is for gear drive or the driven device is such that some mechanical unbalance could occur which might "pound out" a sleeve bearing. Bear in mind that much side pull on a small sleeve bearing causes it to wear so that the inner surface can become egg-shaped. The motor then becomes noisy.

Belt drives with subfractional-horsepower motors can become a problem. Usually the starting torque of the motor may be inadequate for starting a tight, stiff belt. If the belt tension is loosened, it may slip. Small, toothed pulleys and loose "cog belts" can aid in such cases.

Over the range of *fractional-horsepower* ratings both sleeve and ball bearings are in common use. Fan or blower drives which may exhibit little end or side thrust would favor sleeve bearings because of noise. Pump motors, when noise is not too much of a problem, are likely to use ball bearings, because considerable thrust appears in some designs.

It is not uncommon to find in any industry that one company may prefer ball bearings; another with a similar device prefers sleeve-bearing motors.

Until about 25 years ago, nearly all manufacturers of integral-horsepower motors offered a line with either sleeve or ball bearings. More recently, some manufacturers have dropped sleeve bearings, because they find that the greatest part of their business can be satisfied with ball-bearing motors only.

12-9   Lubrication

Mention has been made of the fact that several main classes of grease for ball bearings depend on operating temperature. There are many other greases with slightly different characteristics which form a subject that is gladly ignored here, but nearly every motor manufacturer is aware that it can be a troublesome problem. Some motor buyers desire that ball bearings be packed with one type grease or another (either for good or imaginary reasons). Bear in mind that sealed or shielded bearings are lubricated at the plant of the bearing manufacturer. The result is that a motor builder finds himself stocking identical bearings, identical, that is, in every respect except for grease. Motors supplied to buyer A must always be built with bearings X, and so on.

Or, if the motor construction is such that open bearings are used and the hub reservoirs are partly packed with a certain type of grease, we may find the following condition.

A large buyer of motors uses open bearings in his factory. As a routine matter, his machine maintenance crew periodically greases all equipment, using grease Z in all motors. Is the grease already in the bearings compatible with grease Z? That is the problem, *compatibility*. Many greases mix successfully, others when mixed form a gummy mess with unsatisfactory lubricating qualities.

The same situation exists with oil and sleeve bearings. Porous bearings are impregnated with the specified oil at the factory of the bearing manufacturer. Large buyers of these motors may prefer certain types of oil. A logical criterion is long periods of operation without the oil oxidizing and becoming too gummy to flow. Choices are also made by project engineers with or without adequate data. Again the motor manufacturer stocks identical porous bronze bearings with different oil impregnations.

Perhaps you are the manufacturer of a complex office machine which includes several purchased motors from different manufacturers. Your service man calls regularly on all of your customers and, among other duties, he oils the motors as well as other moving parts. He uses a certain grade oil. Is it compatible with oil already in the motors? If not, another oil and impregnated bearing may be added to the motor manufacturer's inventory.

All of this represents only a minor problem. But as we find out more about the motor industry we will see, if the reader has not already, that it is made up of a multitude of problems, which lead to the often-heard complaint among the industry's personnel. "Certainly there *must* be some easier way of making a living."

Chapter 13

Standardizing Groups, NEMA and Underwriters Laboratories

13-1 Standards of the Industry

The designer of electric motors and the designer of products which use electric motors are influenced by a number of standardizing groups and available published standards.

Of these (unless one is doing business with some branch of the government which has whole libraries of specifications), the most important are the Standards of NEMA (National Electrical Manufacturers Association) and the Underwriter's Laboratories.

13-2 The NEMA Organization

NEMA is the National Electrical Manufacturers Association, a non-profit trade organization supported by manufacturers of electric apparatus and supplies. Its standards are adopted in the public interest and are designed to eliminate misunderstandings between the manufacturer and the purchaser and to assist the purchaser in selecting and obtaining the proper product for his particular need.

13-3 NEMA Standards

NEMA Standards have been developed through continual consultation among manufacturers, designers, users, and national engineering societies. A NEMA Standard defines a product, process, or procedure with reference to one or more of the following: nomenclature, composition, con-

216

struction, dimensions, tolerance, safety, operating characteristics, performance, quality, rating, testing, and the service for which designed. Existence of a NEMA Standard does not in any respect preclude any member or nonmember from manufacturing or selling products not conforming to the standard.

NEMA Standards for motors cover frame sizes and dimensions, horsepower ratings, service factors, temperature rises, and performance characteristics. The benefits such standards provide are greater availability, more convenience in use, sounder basis for accurate comparison, prompter repair service, and shorter delivery time.

13-4 Standardized Motor Frames

Before the NEMA Standards, a manufacturer of, say, machine tools might have purchased a 5-hp 1,725-rpm motor from motor manufacturer *A*. If for any reason he decided to buy from *B*, he might have found that the motor was larger than that of his previous source, had a larger shaft, which was higher above the base, and perhaps the holes in the feet followed a different pattern. In short, he could not mount motor *B* in the same location in which motor *A* had operated.

The first chore to which NEMA applied itself in the 1930's was the standardization of frame sizes, at least to the extent that affected interchangeability. Out of this came a group of standardized NEMA frames applying to 1 hp and above. The frames were numbered from 203 to 505. Standard horsepower and speed ratings were assigned to these.

Table 13-1 lists the frames and their important dimensions for what are called fractional-horsepower sizes, 42 through 66, and integral-horsepower ratings from 143 through 505. In the smaller frames with two digits the frame numbers should be divided by 16 to give the D dimension. This is the height of the center of the shaft above the base and might be called the significant number in this system. Thus for frame 48 the D dimension is 3 in. This, in turn, approximately fixes the motor diameter (less than 6 in.).

For the integral-horsepower frames with three digits, the two left-hand digits should be divided by 4 to give the D dimension. Thus frame 324 will have a D dimension of 8 in. and its diameter will be less than 16 in.

Note, for instance, there are both 324 and 326 frames. The higher number implies a longer frame and a greater F dimension. However, the right-hand digits are merely relative; they are not significant of anything we can measure.

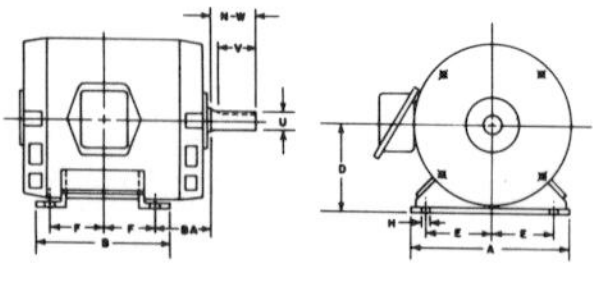

Table 13-1 Dimensions for Foot-Mounted AC Motors*

Frame Number	Dimensions (in.)												
	A max	B max	D†	E	F	BA	H	N-W‡	U‡	V‡ min	Width‡	Key Thickness‡	Length‡
42			$2\frac{5}{8}$	$1\frac{3}{4}$	$\frac{27}{32}$	$2\frac{1}{16}$	$\frac{9}{32}$ slot	$1\frac{1}{8}$	$\frac{3}{8}$			$\frac{3}{64}$ flat	
48			3	$2\frac{1}{8}$	$1\frac{3}{8}$	$2\frac{1}{2}$	$\frac{11}{32}$ slot	$1\frac{1}{2}$	$\frac{1}{2}$			$\frac{3}{64}$ flat	
48H			3	$2\frac{1}{8}$	$2\frac{3}{8}$	$2\frac{1}{2}$	$\frac{11}{32}$ slot	$1\frac{1}{2}$	$\frac{1}{2}$			$\frac{3}{64}$ flat	
56			$3\frac{1}{2}$	$2\frac{7}{16}$	$1\frac{1}{2}$	$2\frac{3}{4}$	$\frac{11}{32}$ slot	$1\frac{7}{8}$	$\frac{5}{8}$		$\frac{3}{16}$	$\frac{3}{16}$	$1\frac{3}{8}$§
56H			$3\frac{1}{2}$	$2\frac{7}{16}$	$2\frac{1}{2}$	$2\frac{3}{4}$	$\frac{11}{32}$ slot	$1\frac{7}{8}$	$\frac{5}{8}$		$\frac{3}{16}$	$\frac{3}{16}$	$1\frac{3}{8}$§
66			$4\frac{1}{8}$	$2\frac{15}{16}$	$2\frac{1}{2}$	$3\frac{1}{8}$	$\frac{13}{32}$ slot	$2\frac{1}{4}$	$\frac{3}{4}$		$\frac{3}{16}$	$\frac{3}{16}$	$1\frac{7}{8}$§
143, T	7	6	$3\frac{1}{2}$	$2\frac{3}{4}$	2	$2\frac{1}{4}$	$\frac{11}{32}$	2, $2\frac{1}{4}$	$\frac{3}{4}$, $\frac{7}{8}$	$1\frac{3}{4}$, 2	$\frac{3}{16}$, $\frac{3}{16}$	$\frac{3}{16}$, $\frac{3}{16}$	$1\frac{3}{8}$, $1\frac{3}{8}$
145, T	7	6	$3\frac{1}{2}$	$2\frac{3}{4}$	$2\frac{1}{2}$	$2\frac{1}{4}$	$\frac{11}{32}$	2, $2\frac{1}{4}$	$\frac{3}{4}$, $\frac{7}{8}$	$1\frac{3}{4}$, 2	$\frac{3}{16}$, $\frac{3}{16}$	$\frac{3}{16}$, $\frac{3}{16}$	$1\frac{3}{8}$, $1\frac{3}{8}$
182, T	9	$6\frac{1}{2}$	$4\frac{1}{2}$	$3\frac{3}{4}$	$2\frac{1}{4}$	$2\frac{3}{4}$	$\frac{13}{32}$	$2\frac{1}{4}$, $2\frac{3}{4}$	$\frac{7}{8}$, $1\frac{1}{8}$	2, $2\frac{1}{2}$	$\frac{3}{16}$, $\frac{1}{4}$	$\frac{3}{16}$, $\frac{1}{4}$	$1\frac{3}{8}$, $1\frac{3}{4}$
184, T	9	$7\frac{1}{2}$	$4\frac{1}{2}$	$3\frac{3}{4}$	$2\frac{3}{4}$	$2\frac{3}{4}$	$\frac{13}{32}$	$2\frac{1}{4}$, $2\frac{3}{4}$	$\frac{7}{8}$, $1\frac{1}{8}$	2, $2\frac{1}{2}$	$\frac{3}{16}$, $\frac{1}{4}$	$\frac{3}{16}$, $\frac{1}{4}$	$1\frac{3}{8}$, $1\frac{3}{4}$
213, T	$10\frac{1}{2}$	$7\frac{1}{2}$	$5\frac{1}{4}$	$4\frac{1}{4}$	$2\frac{3}{4}$	$3\frac{1}{2}$	$\frac{13}{32}$	3, $3\frac{3}{8}$	$1\frac{1}{8}$, $1\frac{3}{8}$	$2\frac{3}{4}$, $3\frac{1}{8}$	$\frac{1}{4}$, $\frac{5}{16}$	$\frac{1}{4}$, $\frac{5}{16}$	2, $2\frac{3}{8}$
215, T	$10\frac{1}{2}$	9	$5\frac{1}{4}$	$4\frac{1}{4}$	$3\frac{1}{2}$	$3\frac{1}{2}$	$\frac{13}{32}$	3, $3\frac{3}{8}$	$1\frac{1}{8}$, $1\frac{3}{8}$	$2\frac{3}{4}$, $3\frac{1}{8}$	$\frac{1}{4}$, $\frac{5}{16}$	$\frac{1}{4}$, $\frac{5}{16}$	2, $2\frac{3}{8}$
254U, T	$12\frac{1}{2}$	$10\frac{3}{4}$	$6\frac{1}{4}$	5	$4\frac{1}{8}$	$4\frac{1}{4}$	$\frac{17}{32}$	$3\frac{3}{4}$, 4	$1\frac{3}{8}$, $1\frac{5}{8}$	$3\frac{1}{2}$, $3\frac{3}{4}$	$\frac{5}{16}$, $\frac{3}{8}$	$\frac{5}{16}$, $\frac{3}{8}$	$2\frac{3}{4}$, $2\frac{7}{8}$
256U, T	$12\frac{1}{2}$	$12\frac{1}{2}$	$6\frac{1}{4}$	5	5	$4\frac{1}{4}$	$\frac{17}{32}$	$3\frac{3}{4}$, 4	$1\frac{3}{8}$, $1\frac{5}{8}$	$3\frac{1}{2}$, $3\frac{3}{4}$	$\frac{5}{16}$, $\frac{3}{8}$	$\frac{5}{16}$, $\frac{3}{8}$	$2\frac{3}{4}$, $2\frac{7}{8}$
284U, T	14	$12\frac{1}{2}$	7	$5\frac{1}{2}$	$4\frac{3}{4}$	$4\frac{3}{4}$	$\frac{17}{32}$	$4\frac{7}{8}$, $4\frac{5}{8}$	$1\frac{5}{8}$, $1\frac{7}{8}$	$4\frac{5}{8}$, $4\frac{3}{8}$	$\frac{3}{8}$, $\frac{1}{2}$	$\frac{3}{8}$, $\frac{1}{2}$	$3\frac{3}{4}$, $3\frac{1}{4}$
284TS	14	$12\frac{1}{2}$	7	$5\frac{1}{2}$	$4\frac{3}{4}$	$4\frac{3}{4}$	$\frac{17}{32}$	$3\frac{1}{4}$	$1\frac{5}{8}$	3	$\frac{3}{8}$	$\frac{3}{8}$	$1\frac{7}{8}$
286U, T	14	14	7	$5\frac{1}{2}$	$5\frac{1}{2}$	$4\frac{3}{4}$	$\frac{17}{32}$	$4\frac{7}{8}$, $4\frac{5}{8}$	$1\frac{5}{8}$, $1\frac{7}{8}$	$4\frac{5}{8}$, $4\frac{3}{8}$	$\frac{3}{8}$, $\frac{1}{2}$	$\frac{3}{8}$, $\frac{1}{2}$	$3\frac{3}{4}$, $3\frac{1}{4}$
286TS	14	14	7	$5\frac{1}{2}$	$5\frac{1}{2}$	$4\frac{3}{4}$	$\frac{17}{32}$	$3\frac{1}{4}$	$1\frac{5}{8}$	3	$\frac{3}{8}$	$\frac{3}{8}$	$1\frac{7}{8}$
324U, T	16	14	8	$6\frac{1}{4}$	$5\frac{1}{4}$	$5\frac{1}{4}$	$\frac{21}{32}$	$5\frac{5}{8}$, $5\frac{1}{4}$	$1\frac{7}{8}$, $2\frac{1}{8}$	$5\frac{3}{8}$, 5	$\frac{1}{2}$, $\frac{1}{2}$	$\frac{1}{2}$, $\frac{1}{2}$	$4\frac{1}{4}$, $3\frac{7}{8}$
324S, TS	16	14	8	$6\frac{1}{4}$	$5\frac{1}{4}$	$5\frac{1}{4}$	$\frac{21}{32}$	$3\frac{1}{4}$, $3\frac{3}{4}$	$1\frac{5}{8}$, $1\frac{7}{8}$	3, $3\frac{1}{2}$	$\frac{3}{8}$, $\frac{1}{2}$	$\frac{3}{8}$, $\frac{1}{2}$	$1\frac{7}{8}$, 2
326U, T	16	$15\frac{1}{2}$	8	$6\frac{1}{4}$	6	$5\frac{1}{4}$	$\frac{21}{32}$	$5\frac{5}{8}$, $5\frac{1}{4}$	$1\frac{7}{8}$, $2\frac{1}{8}$	$5\frac{3}{8}$, 5	$\frac{1}{2}$, $\frac{1}{2}$	$\frac{1}{2}$, $\frac{1}{2}$	$4\frac{1}{4}$, $3\frac{7}{8}$
326S, TS	16	$15\frac{1}{2}$	8	$6\frac{1}{4}$	6	$5\frac{1}{4}$	$\frac{21}{32}$	$3\frac{1}{4}$, $3\frac{3}{4}$	$1\frac{5}{8}$, $1\frac{7}{8}$	3, $3\frac{1}{2}$	$\frac{3}{8}$, $\frac{1}{2}$	$\frac{3}{8}$, $\frac{1}{2}$	$1\frac{7}{8}$, 2
364U, T	18	$15\frac{1}{4}$	9	7	$5\frac{5}{8}$	$5\frac{7}{8}$	$\frac{21}{32}$	$6\frac{3}{8}$, $5\frac{7}{8}$	$2\frac{1}{8}$, $2\frac{3}{8}$	$6\frac{1}{8}$, $5\frac{5}{8}$	$\frac{1}{2}$, $\frac{5}{8}$	$\frac{1}{2}$, $\frac{5}{8}$	5, $4\frac{1}{4}$
364US, TS	18	$15\frac{1}{4}$	9	7	$5\frac{5}{8}$	$5\frac{7}{8}$	$\frac{21}{32}$	$3\frac{3}{4}$, $3\frac{7}{8}$	$1\frac{7}{8}$, $1\frac{7}{8}$	$3\frac{1}{2}$, $3\frac{1}{2}$	$\frac{1}{2}$, $\frac{1}{2}$	$\frac{1}{2}$, $\frac{1}{2}$	2, 2
365U, T	18	$16\frac{1}{4}$	9	7	$6\frac{1}{8}$	$5\frac{7}{8}$	$\frac{21}{32}$	$6\frac{3}{8}$, $5\frac{7}{8}$	$2\frac{1}{8}$, $2\frac{3}{8}$	$6\frac{1}{8}$, $5\frac{5}{8}$	$\frac{1}{2}$, $\frac{5}{8}$	$\frac{1}{2}$, $\frac{5}{8}$	5, $4\frac{1}{4}$
365US, TS	18	$16\frac{1}{4}$	9	7	$6\frac{1}{8}$	$5\frac{7}{8}$	$\frac{21}{32}$	$3\frac{3}{4}$, $3\frac{7}{8}$	$1\frac{7}{8}$, $1\frac{7}{8}$	$3\frac{1}{2}$, $3\frac{1}{2}$	$\frac{1}{2}$, $\frac{1}{2}$	$\frac{1}{2}$, $\frac{1}{2}$	2, 2
404U, T	20	$16\frac{1}{4}$	10	8	$6\frac{1}{8}$	$6\frac{5}{8}$	$\frac{13}{16}$	$7\frac{1}{8}$, $7\frac{1}{4}$	$2\frac{3}{8}$, $2\frac{7}{8}$	$6\frac{7}{8}$, 7	$\frac{5}{8}$, $\frac{3}{4}$	$\frac{5}{8}$, $\frac{3}{4}$	$5\frac{1}{2}$, $5\frac{5}{8}$
404US, TS	20	$16\frac{1}{4}$	10	8	$6\frac{1}{8}$	$6\frac{5}{8}$	$\frac{13}{16}$	$4\frac{1}{4}$, $4\frac{1}{4}$	$2\frac{1}{8}$, $2\frac{1}{8}$	4, 4	$\frac{1}{2}$, $\frac{1}{2}$	$\frac{1}{2}$, $\frac{1}{2}$	$2\frac{3}{4}$, $2\frac{3}{4}$
405U, T	20	$17\frac{3}{4}$	10	8	$6\frac{7}{8}$	$6\frac{5}{8}$	$\frac{13}{16}$	$7\frac{1}{8}$, $7\frac{1}{4}$	$2\frac{3}{8}$, $2\frac{7}{8}$	$6\frac{7}{8}$, 7	$\frac{5}{8}$, $\frac{3}{4}$	$\frac{5}{8}$, $\frac{3}{4}$	$5\frac{1}{2}$, $5\frac{5}{8}$
405US, TS	20	$17\frac{3}{4}$	10	8	$6\frac{7}{8}$	$6\frac{5}{8}$	$\frac{13}{16}$	$4\frac{1}{4}$, $4\frac{1}{4}$	$2\frac{1}{8}$, $2\frac{1}{8}$	4, 4	$\frac{1}{2}$, $\frac{1}{2}$	$\frac{1}{2}$, $\frac{1}{2}$	$2\frac{3}{4}$, $2\frac{3}{4}$
444U, T	22	$18\frac{1}{2}$	11	9	$7\frac{1}{4}$	$7\frac{1}{2}$	$\frac{13}{16}$	$8\frac{5}{8}$, $8\frac{1}{2}$	$2\frac{7}{8}$, $3\frac{3}{8}$	$8\frac{3}{8}$, $8\frac{1}{4}$	$\frac{3}{4}$, $\frac{7}{8}$	$\frac{3}{4}$, $\frac{7}{8}$	7, $6\frac{7}{8}$
444US, TS	22	$18\frac{1}{2}$	11	9	$7\frac{1}{4}$	$7\frac{1}{2}$	$\frac{13}{16}$	$4\frac{1}{4}$, $4\frac{3}{4}$	$2\frac{1}{8}$, $2\frac{3}{8}$	4, $4\frac{1}{2}$	$\frac{1}{2}$, $\frac{5}{8}$	$\frac{1}{2}$, $\frac{5}{8}$	$2\frac{3}{4}$, 3
445U, T	22	$20\frac{1}{2}$	11	9	$8\frac{1}{4}$	$7\frac{1}{2}$	$\frac{13}{16}$	$8\frac{5}{8}$, $8\frac{1}{2}$	$2\frac{7}{8}$, $3\frac{3}{8}$	$8\frac{3}{8}$, $8\frac{1}{4}$	$\frac{3}{4}$, $\frac{7}{8}$	$\frac{3}{4}$, $\frac{7}{8}$	7, $6\frac{7}{8}$
445US, TS	22	$20\frac{1}{2}$	11	9	$8\frac{1}{4}$	$7\frac{1}{2}$	$\frac{13}{16}$	$4\frac{1}{4}$, $4\frac{3}{4}$	$2\frac{1}{8}$, $2\frac{3}{8}$	4, $4\frac{1}{2}$	$\frac{1}{2}$, $\frac{5}{8}$	$\frac{1}{2}$, $\frac{5}{8}$	$2\frac{3}{4}$, 3
504U	25	21	$12\frac{1}{2}$	10	8	$8\frac{1}{2}$	$\frac{15}{16}$	$8\frac{5}{8}$	$2\frac{7}{8}$	$8\frac{3}{8}$	$\frac{3}{4}$	$\frac{3}{4}$	$7\frac{1}{4}$
504S	25	21	$12\frac{1}{2}$	10	8	$8\frac{1}{2}$	$\frac{15}{16}$	$4\frac{1}{4}$	$2\frac{1}{8}$	4	$\frac{1}{2}$	$\frac{1}{2}$	$2\frac{3}{4}$
505	25	23	$12\frac{1}{2}$	10	9	$8\frac{1}{2}$	$\frac{15}{16}$	$8\frac{5}{8}$	$2\frac{7}{8}$	$8\frac{3}{8}$	$\frac{3}{4}$	$\frac{3}{4}$	$7\frac{1}{4}$
505S	25	23	$12\frac{1}{2}$	10	9	$8\frac{1}{2}$	$\frac{15}{16}$	$4\frac{1}{4}$	$2\frac{1}{8}$	4	$\frac{1}{2}$	$\frac{1}{2}$	$2\frac{3}{4}$

* For additional dimensions of these motors with a standardized face mounting, see Chapter 15. Adapted from NEMA Standards MG 1-11.31 and 11.31a.

† Dimension D will never be greater than the values listed, but it may be less so that shims are usually required for coupled or geared machines. When exact dimension is required, shims up to $\frac{1}{32}$ in. may be necessary on frame sizes whose dimension D is 8 in. and less; on larger frames, shims up to $\frac{1}{16}$ in. may be necessary.

‡ Second value, where present, is for rerated T frames. Values for frames 143T through 326TS are final; values for 364T through 445TS are tentative.

§ Effective length of keyway.

13-5   Horsepower Assignments

The next step in motor standardization involved assigning a speed and horsepower rating to a given frame size. Thus a 5-hp 4-pole motor as a standard general-purpose product would be built in frame 215. This was done only for ratings of 1 hp and above.

Since the original standardization the industry has gone through a series of rerating programs, always moving the frame size down for a given horsepower. In this process they abandoned some of the original frames (203, 204, 224, 225) and developed new ones. Frame series 180 and 210 are relatively new, with the 140 series barely in production in 1966.

Refer to Table 13-2. As the industry is now undergoing another rerating program, it will be noted that two frames are apparently assigned to each horsepower. Consider the 15-hp 3,600-rpm motor. For the past 10 years or so it as produced in frame 256U. It is currently obtainable in 215T. (The newer frames all carry the suffix T. The suffixes U and S pertain to shaft modifications.)

A graphical illustration of what has happened in integral-horsepower motor sizes is shown in Figure 13-1. The recent reduction in frame size is brought about mainly by operation at higher temperatures. This involves class B insulating materials versus class A, which will be explained more fully in the chapter that follows.

13-6   Minimum Performance Standards

The next step in this standardization involved minimum values of performance. A user of a 5-hp motor might require occasional overloads, say, $1\frac{1}{2}$ times rated load. Maximum torque developed by the motor should then be well over 200%. Or perhaps the user required a high starting (locked-rotor) torque. Two motor manufacturers might supply good 5-hp motors, but one might be lower in maximum torque and starting torque than the other. Hence the need for establishing *minimum* standards for these values, so that the buyer could depend on interchangeability of performance. As time went on, it was recognized that some users required especially high values of starting torque or high slip, and so various standards of performance were established. Thus the motor buyer has his choice of NEMA designs A, B, C, D, and F.

The various values of expected torques, both locked rotor and breakdown, for various ratings and types are tabulated in NEMA Standards. A few typically contrasted values are shown in Figures 13-2 and 13-3.

Table 13-2 Frame Assignments for Continuous-Duty 60-Cycle Polyphase Motors

Rating (hp)	3600 Open	3600 TEFC	1800 Open	1800 TEFC	1200 Open and TEFC	900 Open and TEFC	Frame Series
½	NA	NA	NA	NA	NA	143T/182	140T
¾	NA	NA	NA	NA	143T/182	145T/184	
1	NA	NA	143T/182	143T/182	145T/184	182T/213	180T
1½	143T/182	143T/182	145T/184	145T/184	182T/184	184T/213	
2	145T/184	145T/184	145T/184	145T/184	184T/213	213T/213	210T
3	145T/184	182T/184	182T/213	182T/213	213T/215	215T/254U	
5	182T/213	184T/213	184T/215	184T/215	215T/254U	254T/256U	250T
7½	184T/215	213T/215	213T/254U	213T/254U	254T/256U	256T/284U	
10	213T/254U	215T/254U	215T/256U	215T/254U	256T/284U	284T/286U	280T
15	215T/256U	254T/256U	254T/284U	254T/284U	284T/324U	286T/326U	
20	254T/284U	256T/284U	256T/286U	256T/286U	286T/326U	324T/364U	320T
25	256T/286U	284TS/286U	284T/324U	284T/324U	324T/364U	326T/365U	
30	284TS/324S	286TS/324S	286T/326U	286T/326U	326T/365U	364T/404U	360T
40	286TS/326S	324TS/326S	324T/364U	324T/364U	364T/404U	365T/405U	
50	324TS/364US	326TS/364US	326T/365US	326T/365US	365T/405U	404T/444U	400T
60	326TS/365US	364TS/365US	364TS/404US	364TS/404US	404T/444U	405T/445U	
75	364TS/404US	365TS/404US	365TS/405US	365TS/405US	405T/445U	444T/NA	440T
100	365TS/405US	405TS/405US	404TS/444US	405TS/444US	444T/NA	445T/NA	
125	404TS/444US	444TS/444US	405TS/445US	444TS/445US	445T/NA	NA	
150	405TS/445US	445TS/445US	444TS/NA	445TS/NA	NA	NA	
200	444TS/NA	NA	445TS/NA	NA	NA	NA	
250	445TS/NA	NA	NA	NA	NA	NA	

Adapted from NEMA Standards MG 1-1302, 13.02a, 13.00, and 13.06a. New Class B frame assignments are shown in boldface; old Class A frame assignments in lightface. NA = Not assigned.

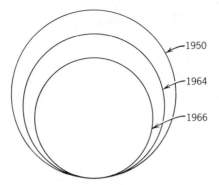

Figure 13-1. Relative diameters of a 7½ hp, 1725 rpm polyphase motor reflect NEMA standards in recent years. The 1966 size utilizes higher temperature insulation and is allowed a higher temperature rise than previous motors.

Note the high-slip motors of Figure 13-3 which were discussed in an example of Chapter 5.

13-7   Fractional-Horsepower Standards

Although dimensions are standardized down to frame 42, horsepower assignments have been left up to the manufacturer's choice on the smaller motors. It will be noted that frames below about 5 in. in diameter are not covered dimensionally by NEMA Standards,[1] although they are built in great varieties and numbers. The fact that they are not NEMA frames is a common misunderstanding on the part of product designers.

13-8   Basis of Horsepower Ratings for Single-Phase Motors

In single-phase motors the horsepower, surprisingly enough, is fixed by the maximum or breakdown torque. The uninitiated might consider that a ½-hp 1725-rpm motor is one capable of operating continuously at ½-hp loads without excessive temperature rise. This is only a part of the picture. In accordance with NEMA it is also defined as a motor which has a breakdown torque of between 40.5 and 58.0 oz-ft.

Complete tables of values for all single-phase motors rated from 1 mhp to 10 hp are given in the NEMA Standards.

[1] This excludes the partial dimensioning system set up for some instrument motors.

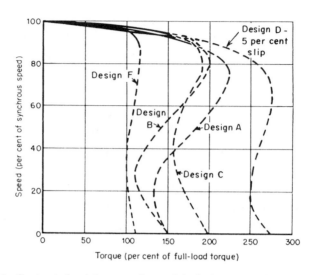

Figure 13-2. Contrasted minimum values of locked rotor and maximum torques for various standardized designs. (From *Electric Motors*, Reference Issue of *Machine Design*.)

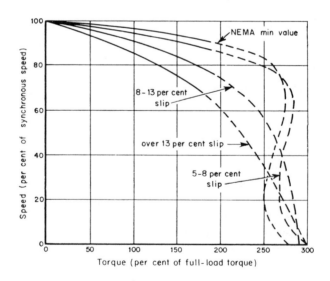

Figure 13-3. Standardized motors are available for various degrees of full-load slip. As explained in Chapter 5, this dropoff in speed with load is an important factor in the application of certain motors. (From *Electric Motors*, Reference Issue of *Machine Design*.)

13-9   Service Factor

A general-purpose motor is expected to be operated at overload without unsafe operating temperatures. Efficiency, power factor, and speed will be different from rated values.

Maximum overload values are obtained by multiplying the rated horsepower by the service factor, as given in Table 13-3.

13-10   Locked-Rotor Current and Kilovolt-Amperes

The locked-rotor current of a motor is the large inrush of current when the motor is connected to its supply. It can and does cause a dip in the supply voltage. Excessive currents are to be avoided and, therefore, maximum values have been established as measured at rated voltage and frequency. These are given for both the old and the rerated motors in Table 13-4.

These values are important in determining the wiring from motor to supply, overload protection, etc. NEMA, in line with the National Electrical Code, requires that a code letter be used on the nameplate of all motors $\frac{1}{20}$ hp or over, which indicates approximately the locked-rotor current. Code letters are given in Table 13-5.

Table 13-3

Horsepower	Service Factor, AC Motors
$\frac{1}{20}$	1.4
$\frac{1}{12}$	1.4
$\frac{1}{8}$	1.4
$\frac{1}{6}$	1.35
$\frac{1}{4}$	1.35
$\frac{1}{3}$	1.35
$\frac{1}{2}$	1.25*
$\frac{3}{4}$	1.25*
1	1.25*
$1\frac{1}{2}$	1.20*
2	1.20*
3 and larger	1.15*

* In the case of polyphase squirrel-cage integral-horsepower motors, these service factors apply only to design A, B, C, and F motors (MG1-12.44 and MG1-14.33).

To calculate the starting current from the code letter,

$$\text{amperes} = \frac{\text{kva/hp (from Table 13-5)} \times \text{hp} \times 1{,}000}{\text{volts} \times (1.73 \text{ for three-phase motor})} \cdot$$
$$(1.00 \text{ for single-phase motor})$$

Thus, for a three-phase motor rated at $7\frac{1}{2}$ hp, 440 v, showing a code letter F,

$$\text{amperes} = \frac{(5.00 \text{ to } 5.59) \times 7\frac{1}{2} \times 1{,}000}{440 \times 1.73}$$
$$= 49.2 \text{ to } 55.$$

13-11   Underwriters' Laboratories

The Underwriters' Laboratories, Inc., is a nonprofit organization sponsored by the American Insurance Association. It operates laboratories

Table 13-4 Locked-Rotor (Starting) Currents for Induction Motors. (*a*) By Present Standards for Three-Phase Motors. (*b*) For Rerated Class *B* Motors in the T Frames. (*c*) For Fractional- and Integral-Horsepower. Single-Phase Motors.

(*a*) Locked-Rotor Current Amperes at 220 v,
Three-Phase Induction Motors*

Horse-power	60-Cycle Motors Designs B, C, and D	50-Cycle Motors, Design B	Horse-power	60-Cycle Motors Designs B, C, and D	Design F	50-Cycle Motors, Design B
1 or less	30	‡	30	435	270	500
$1\frac{1}{2}$	40	40	40	580	360	670
2	50	50	50	725	450	835
3	64	70	60	870	540	1,000
5	92	105	75	1,085	675	1,250
$7\frac{1}{2}$	127	140	100	1,450	900	1,670
10	162	175	125	1,815	1,125	2,090
15	232	255	150	2,170	1,350	2,495
20	290	335	200	2,900	1,800	3,335
25	365	420	. . .	. . . . .	. . . . .	. . . . .

*Locked-rotor current of motors designed for voltages other than 220 v shall be inversely proportional to the voltages.
‡ 28 amperes per horsepower (MG1-12.33).

Table 13-4   (Continued)

(b) Three-Phase, 60-Cycle Motors at 230 v*

Horsepower	Locked-Rotor Current, Amp	
	Designs B, C, and D	Design F
1 or less	30	. . . . .
$1\frac{1}{2}$	40	. . . . .
2	50	. . . . .
3	64	. . . . .
5	92	. . . . .
$7\frac{1}{2}$	127	. . . . .
10	162	. . . . .
15	232	. . . . .
20	290	. . . . .
25	365	. . . . .
30	435	270
40	580	360
50	725	450
60	870	540
75	1,085	675
100	1,450	900
125	1,815	1,125
150	2,170	1,350
200	2,900	1,800

* Locked-rotor current of motors designed for voltages other than 230 volts shall be inversely proportional to the voltages. Note: This suggested Standard for Future Design relates to the rerating of integral-horsepower induction motors. (MG1-12.33a)

for testing devices, systems and materials relative to life, fire and casualty hazards. Although its activities cover a broad field, our concern is with their standards for motors, motor components, and motor-driven equipment. The basic idea is that the device should not constitute a hazard to life nor a cause of fire.

13-12   Procedure

The usual practice is for the manufacturer of anything from a meat saw to a hair drier to submit his product to the Underwriters' Laboratories for testing and, hopefully, final listing as approved in their bulle-

Table 13-4   (Continued)

(c) Fractional-Horsepower
Two-, Four-, Six-, and Eight-Pole 60-Cycle Motors Single-Phase

| | Locked-Rotor Current, amp | | | |
| | 115 v | | 230 v | |
Horsepower	Design O	Design N	Design O	Design N
$\frac{1}{6}$ and smaller	50	20	25	12
$\frac{1}{4}$	50	26	25	15
$\frac{1}{3}$	50	31	25	18
$\frac{1}{2}$	50	45	25	25
$\frac{3}{4}$	. . .	61	. . .	35

The locked-rotor currents of single-phase general-purpose fractional-horsepower motors shall not exceed the values for design N motors. (MG1-12.32)

Integral-Horsepower, Designs L and M

| | Locked-Rotor Current, amp | | |
| | Design L Motors | | Design M Motors |
Horsepower	115 v	230 v	230 v
1	70	35	. . .
$1\frac{1}{2}$	. . .	50	40
2	. . .	65	50
3	. . .	90	70
5	. . .	135	100
$7\frac{1}{2}$	. . .	200	150
10	. . .	260	200
15	. . .	390	300
20	. . .	520	400

(MG1-12.34)

tins. It is obvious that their tests on an electric fan might differ greatly from those for a garbage disposer; but our concern is with the tests pertaining to the motor drive.

The device is operated normally and/or under special tests devised by the Laboratories. The temperature rise of the motor is observed under these conditions. On the assumption that in the home or in usual commercial surroundings a reasonable ambient temperature is 25°C, temperature

Table 13-5   Code Letters for Locked-Rotor Kilovolt-Amperes

code letter	locked-rotor kva per horsepower*
A	0– 3.15
B	3.15– 3.55
C	3.55– 4.0
D	4.0 – 4.5
E	4.5 – 5.0
F	5.0 – 5.6
G	5.6 – 6.3
H	6.3 – 7.1
J	7.1 – 8.0
K	8.0 – 9.0
L	9.0 –10.0
M	10.0 –11.2
N	11.2 –12.5
P	12.5 –14.0
R	14.0 –16.0
S	16.0 –18.0
T	18.0 –20.0
U	20.0 –22.4
V	22.4 –and up

* Locked-rotor kva per horsepower range includes the lower figure up to, but not including, the higher figure. For example, 3.14 is letter A and 3.15 is letter B.   [MG1-10.38]

rises of 75°C are permitted. Bear in mind that this same motor may have been used on an entirely different device which has already been listed by the Laboratories. But this may mean very little since the present appliance may be so built as to interfere with the motor ventilation, or may be subject to overloads differing from its previous application. These conditions could cause overheating and potential hazard.

In addition to such tests, the motor construction is checked for clearances between live parts and body as well as to the size of openings through which probes might be extended which could contact live parts. If a thermal protective device is used, the combination is also checked to see that the protector is really effective.

13-13   Motor Listing

One of the least understood aspects of the foregoing procedure is the fact that the manufacturer of the complete device must apply for approval. Motor manufacturers are frequently requested to apply for approval of a certain motor in the belief that when this is obtained, the motor-driven device is also approved.

Actually a motor may be submitted and approved on a limited basis, called a *yellow card* listing. This indicates that the motor has been examined for clearances, mechanical protection, and other characteristics which appear independently of its application. When this identical design is used to drive a device for which approval is requested, some time is saved as a result of the previous investigation.

13-14   Explosionproof Motors

Underwriters' Laboratories also set the standards and do the testing for explosionproof motors to be used in a variety of hazardous locations. These are motors built mechanically strong enough to withstand an explosion within their frame without sparks or flashes spreading to the surrounding atmosphere (see Figure 13-4).

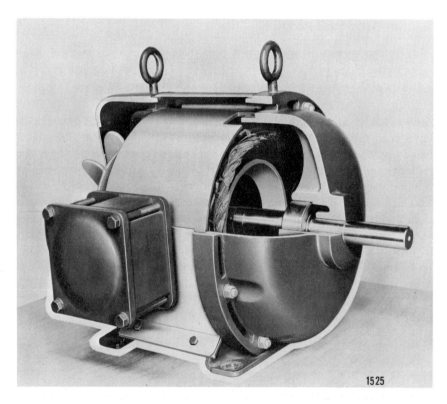

1525

Figure 13-4. Cutaway view of a totally enclosed, fan-cooled, explosionproof motor. Note the overlap of the head fit to the body and long close tolerance clearance between hub and shaft. As an enclosed motor, the windings would overheat except for the air flow maintained between the motor body and outer shell by the external fan. (Courtesy of Robbins & Myers, Inc.)

They are classified by the hazardous gases or dusts to which they might be exposed. The mechanical designs required differ slightly to meet the requirements of the various conditions.

Suppose an electric motor is to be used in a refinery surrounded by highly combustible gases. The theory is that this gas cannot be kept out of the motor and that the motor will ultimately burn out. When it does, the resulting sparks will ignite the gases within the motor and an explosion occurs. The motor must not crack open, nor can any flame extend along the shaft or at any of the fitted seams.

13-15   Policing the Standards

In addition to making the original tests on the usual devices, appliances, etc., as well as on the explosionproof motors, the Laboratories maintain an annual reexamination of its listed products at the plant of the manufacturer. This gets down to the motor manufacturers level if they have yellow-card listing, and very rigorously so if their label service is used on explosionproof motors.

On the other hand, the buyer of an electric motor which is represented as meeting NEMA Standards, both mechanically and electrically, depends on the integrity of the supplier. If it fails to do so, the motor manufacturer has merely risked losing a customer. Failure to comply with Underwriters' Laboratories requirements involves hazards to personnel and property and, hence, follows a more rigorous policy.

Chapter 14

Losses, Temperature Rise, Insulation, Application Example

___

14-1 Motor Losses

In dealing with the magnetic circuit we found that lamination steel, subjected to an alternating flux, displayed hysteresis loss and also eddy current losses. The latter are of two kinds—circulating currents within a single lamination and also between laminations; the latter occurs in spite of the oxide coating built up on the lamination surfaces.

In addition, we naturally expect an I^2R loss from the current in the windings, both in the stator coils and the aluminum bars and end rings of the cage. Friction in the bearings and "windage" loss from the fans complete the list of normal losses (see Chapter 5).

It is of interest to examine the magnitude of these losses. In Figure 14-1 this breakdown is given for a 1-hp four-pole three-phase motor. It is understood, of course, that these losses all appear as heat, and that this heat must be removed from the motor if its windings are to be kept at a safe temperature.

14-2 Temperature Rise

A hot surface dissipates heat at a greater rate than a cool surface. Suppose a 100-w lamp bulb were placed in a coffee can with the lid on. Temperature measured on the outside of the can will go up rapidly at first and, then, gradually level off until a final steady value is reached. At this point the temperature of each square inch of the can surface has reached a value above the surrounding air at which it dissipates heat exactly at the same rate as heat is being supplied to it by the lamp. Put this whole setup into a refrigerator and it will be found that

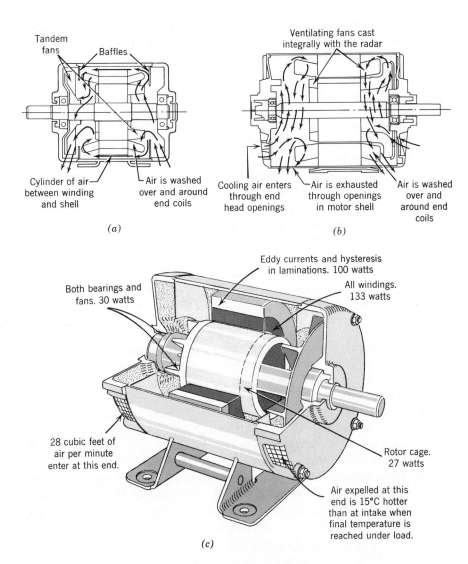

Tandem fans

Baffles

Cylinder of air between winding and shell

Air is washed over and around end coils

(a)

Ventilating fans cast integrally with the radar

Cooling air enters through end head openings

Air is exhausted through openings in motor shell

Air is washed over and around end coils

(b)

Eddy currents and hysteresis in laminations. 100 watts

Both bearings and fans. 30 watts

All windings. 133 watts

28 cubic feet of air per minute enter at this end.

Rotor cage. 27 watts

Air expelled at this end is 15°C hotter than at intake when final temperature is reached under load.

(c)

Figure 14-1. One common ventilating scheme for motors (a) takes air in at one end of the motor and expels it at the other. A second construction (b) takes in air at each end head, circulates it over the end coils and exhausts it through openings in the body. Air does not flow from end to end in this method. Typical losses of of 1 hp, 1725 rpm, three-phase motor are shown in (c). They total 290 w at rated load and voltage.

the final temperature is less but that the rise above the surrounding air is exactly the same. The point is that it is the rise above the surrounding air (ambient) which determines the dissipation rate.

14-3   Motor Ventilation

Return now to the 1-hp motor with the losses shown in Figure 14-1. This is built as an open ventilated motor with a large fan which brings in air at one end head, "washes" it over the end coils and the lamination stack, and out through the other end. Measurements show that this air flow is 28 ft³/min, and when the motor is loaded until a final operating temperature is reached, the output air is 15°C hotter than the input. (All temperature measurements in electric motors are to the Centigrade rather than the more familiar Fahrenheit scale. A conversion table can be found in the Appendix.)

Raising air temperature to a certain level requires a definite amount of energy. We know that it takes 8.4 w to increase 1 ft³ of air by 15°C (of course, this varies somewhat with the density and humidity).

$$28 \text{ ft}^3/\text{min} \times 8.4 \text{ w} = 235 \text{ w.}$$

Thus out of 290-w loss to be dissipated from the motor, 235 are carried out by the ventilating system. The balance is dissipated from the surface of the end heads and motor body.

Consider the motor as a cylinder with a diameter of 7.625 in. and a length of 10.25 in. The cylindrical surface then has an area of 246 in.². Each end head has an area of 45.6 in.². (Of course, these measurements only approximate the actual motor, but we are more interested in illustrating some principles than in developing a rigorous calculation.)

The temperature rise was measured at a number of spots all over the body and heads during the course of a load run on the motor. Temperatures varied considerably but the following appear to be reasonable averages.

End head near fan	15°C rise
Opposite end head	10°C rise
Body	22°C rise

To determine how many watts in heat are dissipated from these surfaces, use the curves of Figure 14-2 for grey lacquer finish.

$$0.00825 \times 15°C \times 45.6 \text{ in.}^2 = 5.63 \text{ w}$$
$$0.00805 \times 10°C \times 45.6 \text{ in.}^2 = 3.67 \text{ w}$$
$$0.00855 \times 22°C \times 246 \text{ in.}^2 = 46.50 \text{ w}$$
$$\text{Total} \qquad\qquad 55.80 \text{ w}$$

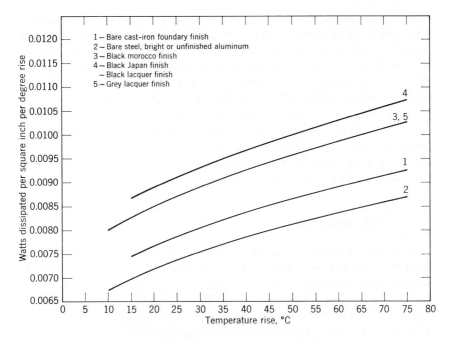

Figure 14-2. Heat dissipation from surfaces.

These watts plus those carried off by the air should equal the 290-w loss.

14-4   Total Enclosure

Let us assume now that the foregoing motor is built in a frame which is totally enclosed. That is, all openings are closed and there is no free access of air from the outside to the inside of the motor. Let us assume also that we could permit a maximum rise on the outside of the motor frame of 75°C. This is rather an impractically high value, for with a room temperature of 25°C (77°F) this would mean that the frame would show a final temperature of 100°C, which is at the boiling point of water. Furthermore, it will be assumed that the entire frame will be at this uniform temperature although that is never achieved in practice. Nevertheless, even going to these extremes it is of interest to determine what will happen under these circumstances.

Again using the curves of Figure 14-2,

$$0.01025 \times 75°C \times (246 + 45.6 + 45.6) = 260 \text{ w.}$$

Since the motor has a full-load loss of 290 w, it can be seen that, even with a uniform temperature rise of 75°C on the motor body, enough heat cannot be dissipated. But if the motor surface is at this temperature, the winding, which is the critical item, will probably be hotter by about 20°. Final operating temperature of the winding, including the ambient, would be above 120°C. This may or may not be practicable.

The points to be kept in mind are that (a) in the normal motor much of the heat is carried off by the ventilating system and anything which interferes with this air flow will tend to overheat the motor and (b) comparative little heat can be dissipated from a surface even with a considerable rise in temperature above the surrounding air.

14-5 Insulation

As we would naturally expect, the voltage applied to an electric motor must be insulated from the frame or "ground" as it is usually referred to. The wires also must be insulated from each other to prevent short-circuits or leakage of current from one wire to another.

The latter function is accomplished by the covering on the magnet wire and by insulating separators between various coils or layers of wire both in the slots and between the end coils (see Figure 14-3).

Insulation of the voltage to ground is achieved by a layer of material in the slot. This is called the *slot cell*. After the winding is in place, it is held down in the slot by an insulating wedge which also prevents the wire from coming into intimate contact with the steel laminations at the tooth tip.

When the windings are inserted and the connections made, the entire assembly, stator laminations, magnet wire, and attached leads, are dipped in insulating varnish, drained, and baked. This might be repeated several times. Ideally the varnish should penetrate between all of the wires and fill the voids of any porous materials used as insulation. It adds to the insulating qualities and prevents moisture from soaking into dangerous places.

The life of an electric motor depends on the life of the insulation. If it deteriorates to the point of breaking down, naturally the motor is burned out or grounded. Insulation ages, and this aging is accelerated by heat.

14-6 Insulation and Temperature Rise

The allowable temperature rise of a motor winding is fixed by the limiting operating temperature of its insulation. When the motor industry

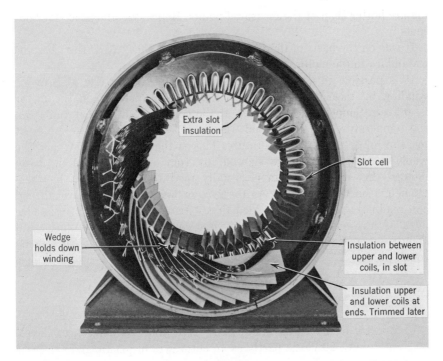

Figure 14-3. Partly wound six-pole, polyphase motor showing some coils and insulation in place. Because these coils lap over one another, this is a "lap" winding in contrast to concentric winding with each coil of a phase and pole group about the same center.

was young, these insulating materials were largely paper, cotton, and enamels or varnishes that deteriorated at much over 100°C.

Allowable temperature rise was set up as follows:

Assumed ambient, 40°C (104°F).
Measured rise on the windings, 40°C.
Allowance for "hot spot" and overload, 25°C.

These add to 105°C, which was set as the upper limit of operation for cotton, paper, etc. At this temperature, insulation is supposed to last for about 20 years.

If we measure temperature rise on the end coils of a motor by thermometers or thermocouples,[1] it is assumed that many parts of the wind-

---

[1] If two dissimilar metals are welded together at a junction and this junction is attached to any surface, a voltage is generated of a magnitude depending on the metals and the temperature of the surface. By reading this voltage we have a measure of the temperature. This is called a *thermocouple*.

ing are hotter than at the measured points. Hence, the allowance for "hot spots."

It was on the basis of the foregoing reasoning that "40°C rise" became a standard in the motor industry for many years.

It is possible to determine the average temperature rise of a motor winding by measuring its resistance. If this is done when the motor is at room temperature, and then at the conclusion of a heat run when the temperature has leveled off, the increase in resistance is an exact measure of the rise in temperature. This is the average throughout the whole winding. Experience showed, by comparing readings on many motors, that rise by resistance usually read 10° higher than by thermometers on the end coils. As the system was established through end coil readings, it was then decided that a temperature rise by resistance of 50°C was equivalent to 40°C by end coil temperature readings, and standards recognize these apparent discrepancies as still being compatible.

14-7 Newer Insulating Material

Many new insulating enamels, films, and laminates have been developed in the past 20 years, partly as an outgrowth of plastic and synthetic fibre development. For their use as insulating materials they, are classified as follows:

Maximum operating temperature 105°C—class A.
Maximum operating temperature 130°C—class B.[2]
Maximum operating temperature 155°C—class F.
Maximum operating temperature 180°C—class H.

A great variety of these insulating materials is available, and we naturally ask, since they are new, how can the user make sure that they will have a probable life of 20 years if operated at their recommended temperatures. A standardized series of accelerated life tests are used. When the results are compared with similar results on known materials of proved longevity, predictions appear to be valid.

14-8 Forty-Degree Rise as a Fetish

For many years when only class A insulation was common, practically all motors were designated as "40°C rise." A motor with a 45 or 50°C

---

[2] Class B insulation is of long standing, using asbestos and mica. These bulky materials have been largely replaced by the newer plastics.

rise was judged as inferior, and many motor buyers took the attitude that, in using such a product, they were in some manner being cheated. In some cases motors have been redesigned at great expense to satisfy a rigorous adherence to this standard by large OEM buyers.

The theory is that for each 8 to 12° over 105°C, the useful life of organic insulating materials is cut in half. This is based on *continuous* operation. Thus, if a motor operated at a higher temperature so that the insulation was subjected to 115°C for 12 hr per day, it would still have a presumed 20-year life. For intermittent service at still higher temperatures, the life span could be even less affected.

Competition in the magnet wire field as well as in all types of insulating materials has had the effect of reducing prices for higher-temperature materials. Thus many motors sold as class A are actually wound with magnet wire with film good for 200°C and have slot cells, wedges, and coil separators rated at 130°C. The only class A materials actually found in these motors might be the tie strings used for binding the end coils (cotton) and the insulation on the leads. This is good business, because it gives the motor an extra factor of safety if misused and it enables the motor manufacturer to simplify his large stock of insulation at little added first cost.

A large factory filled with hundreds of large industrial motors represents a considerable investment. These motors may operate 24 hr per day, and every effort should be made to assure them long life.

But a motor on a business machine or domestic appliance represents a different situation. Its use is usually less strenuous, and it will probably become technologically obsolete or outdated by styling long before the motor life is exhausted.[3]

14-9   Effect of Insulation on Motor Rating

The early magnet wire used in motor windings was coated with enamel and then covered with cotton fibres. The paper insulation in the slot

[3] As can be suspected, the author lacks the usual respect for 40°C rise as a criterion for class A motors. Bear in mind that it arose because of the limitation of 105°C as a final operating temperature for organic insulating materials. This, in turn, was established many years ago by a committee of what was then the AIEE. The group had gathered what data it could, and in a morning meeting had argued as to the relative merits of values from 90 to 115°C. Finally one member said, "Hell, I'm hungry, let's make it 105°C and go to lunch."

One wonders under what temperature limitations the motor industry might have operated these many years if each of the committee members had eaten a hearty breakfast.

was comparatively thick, both for mechanical strength and insulating qualities. As a result, the amount of useful copper that could be placed in a slot was low, perhaps 27% copper and 73% insulation and lost air space.

The development of tough, abrasive resisting enamels for the magnet wire plus thin films for the slot insulation enabled perhaps 35% of the slot area to be filled with useful copper. These films have superior insulating qualities and are tough enough to resist the crushing and tearing that occurs in inserting windings and forming them into place.

If we can put 35% copper in an area that held only 27% before, the resistance and the copper losses might become $\frac{27}{35}$ or 77% of what they were before. Hence a motor manufacturer might be able to design, say, $7\frac{1}{2}$ hp in a motor package, which years ago was rated at 5 hp, and end up with a motor no hotter than the original design.

In addition to this the developments in higher temperature, tough, insulating materials has made it feasible to operate motors with *winding temperature* rises of 80, 105, and 125°C, as shown in Table 14-1. Now a motor frame which a few years ago had a 5-hp rating assigned to it, and then a $7\frac{1}{2}$-hp rating, might now, as a class B insulated motor be designed for 10 hp. It will run hot, but not hotter than the permissible operating temperature of the insulating material. And it is hot because, being in a smaller diameter frame, the fan diameter is too small to force much cooling air through the motor and its square inches of surface for heat dissipation are less.

14-10   Thermal Protectors

A motor is likely to burn out because of dirt blocking off its ventilating passages, frequent starting, some malfunction of the driven device, or just plain overloading. In many cases this burnout can be prevented if a device is provided in the motor which senses the heat and opens the circuit. A frequent form of this device consists of a bimetal disk with two contacts on its rim, which, in turn, connect with two stationary contacts. A smaller heater coil is adjacent to this disk, which carries the motor current. Built into the motor adjacent to the end coils, it senses the heat from the winding and the unusual heat from its own heater coil if the current is abnormal. The bimetal disk heats, snaps from a convex to a concave shape, opening the contacts and disconnecting the motor from the line. When the disk cools, it restores the connections and the motor starts up again. If the trouble is persistent, this cycle continues. The final temperature after this continuous cycling is limited by an Underwriters' Laboratories requirement.

Table 14-1   AC Motor Temperature Limits

	Temperature (°C)				
	Class A		Class B	Class F	Class H
1.0 service factor, dripproof					
Ambient temperature	40	40	40	40	40
Rise by thermometer	40	...	...	...	...
Rise by resistance	...	50	80	105	125
Service-factor margin	10	10	...	...	...
Hot-spot allowance	15	5	10	10	15
Total temperature	105	105	130	155	180
Totally enclosed fan-cooled					
Ambient temperature	40	40	40	40	40
Rise by thermometer	55	...	...	...	...
Rise by resistance	...	60	80	105	125
Hot-spot allowance	10	5	10	10	15
Total temperature	105	105	130	155	180
Totally enclosed nonventilated*					
Ambient temperature	40	40	40	40	40
Rise by thermometer	55	...	85	...	...
Rise by resistance	...	65	...	110	135
Hot-spot allowance	10	0	5	5	5
Total temperature	105	105	130	155	180
Encapsulated					
Ambient temperature			40	40	
Rise by resistance			85	110	
Hot-spot allowance			5	5	
Total temperature			130	155	
1.15 or higher service factor; all motors					
Ambient temperature			40	40	
Rise by resistance†			90	115	
Hot-spot allowance			10	10	
Total temperature			140	165	

* Including all fractional-horsepower totally enclosed motors and fractional-horse-power motors smaller than frame 42.
† At service-factor load.
Adapted from MG 1-12.39 and 12.40.
(From Electric Motors Reference Issue of Machine Design.)

On a motor for, say, oil burner operation, the fact that it is equipped with an *automatic reset* thermal protector may cause no harm. But if applied to a motor for a table saw, overload may stop the motor, the operator starts to investigate, the saw comes back on, and injuries may result. To cover such applications, the bimetal disk can be so designed that it does not close again until pressure is exerted on it by a projecting button. This is the manual reset type (see Figure 14-4).

These manual resets are available in various sizes and, also, instead of using a bimetal disk, a strip is used, resulting in a small capsule-type protector which can be inserted with the motor windings.

Modified forms of this same idea are used with large motors in which the protector is tied to the end coils of the motor and senses only the heat, not the motor current. High temperature does not open the main

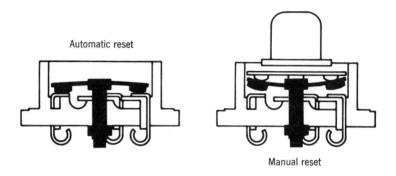

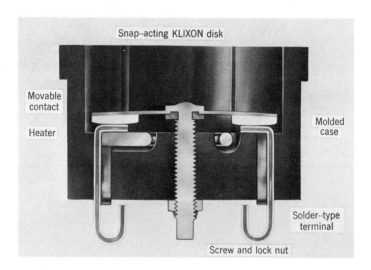

Figure 14-4. One form of thermal overload protector usually mounted in the motor and head adjacent to the windings. Heat from the coil and from the windings causes the bi-metal disk to snap open at a predetermined temperature. When cooled, the contacts may close, depending on the construction. (Courtesy of Metals and Controls Division of Texas Instruments)

line but opens the magnet coil of the main switch and disconnects the motor.

14-11   Applying A Motor

From what has been seen up to this point concerning motor characteristics and ventilation and heating, we are in a position to consider how this information can be applied to the development of a motor-driven device.

A new office machine is in development and has reached the stage wherein the prototype does its required functions and is ready to be matched to a suitable driving motor. A cabinet or case will then be designed to house the entire unit.

During the development, the project engineers had driven the samples by a $\frac{1}{6}$-hp 1725-rpm motor of the split-phase type which happened to be available. From input watts, read on the motor while the device was working, it was suspected that this motor was larger than required. The starting torque of the split-phase motor was measured and found to be 7.2 oz-ft. The voltage was reduced in steps while the motor was connected to the machine, and it was found the machine was still capable of being started when the motor voltage was reduced to 50. From this the project engineer concluded that the motor starting torque could be as low as

$$\left(\frac{50}{115}\right)^2 \times 7.2 = 1.38 \text{ oz-ft.}$$

This calculation was based on the fact that torques vary as the square of the voltage. With this low starting-torque requirement he assumed that a permanent-split-capacitor motor would be adequate, and it offered the possible advantage of quiet operation.

He then requested a motor supplier for a calibrated sample of a nominal 1725-rpm permanent-split-capacitor motor rated at about $\frac{1}{6}$ or $\frac{1}{8}$ hp, open ventilated, with resilient mounting. The motor was to be "calibrated" at 105, 115, and 125 v.

The calibration of a motor can take several forms. Basically, it means that the motor manufacturer makes complete tests measuring speed, amperes input, watt input, and power factor versus horsepower output at several different voltages, and supplies these data with the identical motor on which the tests were made. On the other hand, the calibrated data can take the simple form shown in Figure 14-5, using watt input versus horsepower output.

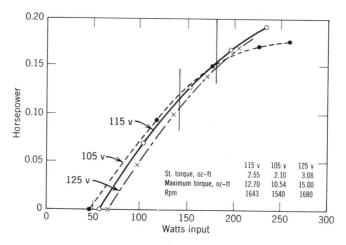

Figure 14-5. Calibration curves for a $\frac{1}{6}$ hp, four-pole permanent split capacitor motor. (6 mfd).

The engineer then used this motor to drive his device and found that, as it went through its operating cycle, the input watts varied from 140 to 180 w at 115 v.

From the curve data this corresponded to 0.112 and 0.155 hp respectively. Losses vary from 56.5 to 64 w. On the basis of its maximum torque of 12.7 oz ft, this could be rated as a $\frac{1}{6}$-hp motor. The motor was also operated at 105 and 125 v and found to be satisfactory, although the reduced speed at 105 v was of some concern. Because the watt input was slightly higher at 125 v, a heat run was made for several hours, operating the device continuously, and a temperature rise of 35°C was noted on the end coils (see Figure 14-6).

The foregoing procedure represents an intelligent approach to the problem of selecting a motor for a device in development. It is not followed universally.

14-12   Cooling the Finished Product

All of the above mentioned testing was done with the equipment completely exposed. The next step involves design of a rough cover or cabinet in which the device would be housed. This, in turn, is often given to

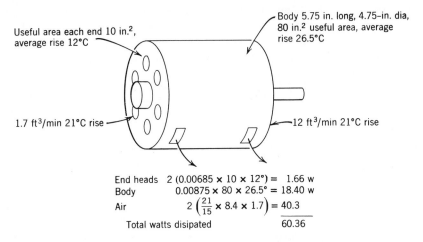

Useful area each end 10 in.², average rise 12°C

Body 5.75 in. long, 4.75-in. dia, 80 in.² useful area, average rise 26.5°C

1.7 ft³/min 21°C rise

12 ft³/min 21°C rise

End heads	$2 (0.00685 \times 10 \times 12°) =$	1.66 w
Body	$0.00875 \times 80 \times 26.5° =$	18.40 w
Air	$2 \left(\frac{21}{15} \times 8.4 \times 1.7\right) =$	40.3
Total watts disipated		60.36

Figure 14-6. This sketch shows the areas and heat dissipation of a $\frac{1}{6}$-hp permanent-split capacitor motor. The capacitor, remotely connected, is ignored. Motor body is gray. The end heads are unfinished die-cast aluminum.

an industrial designer who rounds off the corners and "improves the silhouette." The cabinet may end up with no visible openings in the sides or top except for those which are functional for its operation. Perhaps some openings are provided in the bottom. Ventilating louvers near the motor are useful but are too rarely provided. Industrial designers seem to consider holes and louvers as unsightly.[4]

The losses in this motor average 60 w. Again, if we assume that the air temperature in and out of the cabinet rises 15°C, each cubic foot will carry off 8.4 w. Cooling this unit by air flow alone then requires 7.15 ft³ of air/min. As this is not drawn in and expelled by an fan, nothing but the natural "chimney effect" from the bottom openings to those near the top can permit such circulation of air.

Without this flow the air in the cabinet naturally heats up more, the walls become warm, and their increased temperature increases the heat dissipation from the surface. A final state of equilibrium is reached

[4] We have only to look at the domestic, commercial, and industrial appliances and equipent of 20 years ago and compare them with the modern equivalent to realize the important contribution made by the industrial designer. Nevertheless, they have, on occasion, caused the motor industry many headaches and have been the cause of their clients' paying special prices for motors.

when the air and surface temperature rises are able to dissipate heat equivalent to the 60-w average loss.[5] This could be satisfactory, but, on the other hand, the conditions described too frequently result in a hot motor and a cabinet which feels too warm to touch (see Figure 14-7).

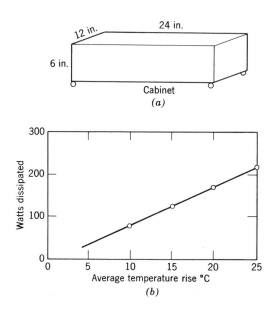

Figure 14-7. Assume that this unidentified motor-driven device is built into the totally enclosed gray lacquered cabinet as shown. The watts that can be dissipated from the total surface is shown in (b). These figures are based on average temperatures but actually they would vary greatly from spot to spot. Note that if only 60 watt motor losses are considered the average temperature rise is 7.5°C. Since all of the work by this motor is apparently performed in the cabinet, the entire input (average 160 watts) appears therein as heat. Under these circumstances the average rise is 19°C. The operating temperature of the motor is increased a like amount. Openings in the cabinet close to the ends of the motor and also in or near the top, would greatly improve these temperatures.

[5] All electric energy finally appears as heat. The mechanical output of an electric motor is always used up in friction, regardless of whether it is pumping water, driving a fan, running a lathe, or what not. In this case, friction in the bearings of the device or that associated with any other moving parts could mean that the entire 140 to 180-w input to the motor might have to be dissipated from the cabinet unless it did some work exterior to the enclosure.

The final indignity occurs when the producer decides that the unit is noisy and lines the inside of the cabinet with sound-deadening material. This acts as a heat insulation and cuts down the dissipation, causing the inside air to become hotter still.

The circumstances described in the foregoing are not exceptional; they occur all too often in product design. The attitude seems to be that if heat problems are ignored they will some how go away. They never do.

Chapter 15

Definite- and Special-Purpose Motors

In a broad sense nearly all electric motors can be grouped into three classifications: general purpose, definite purpose, and special purpose.

The general-purpose motor of fractional horsepower sizes is built in standard NEMA frames with standard torques and starting current. It can be purchased from local repair shops, mail order houses, and possibly the local drug store. The largest business in general-purpose motors is done in the integral-horsepower sizes. Again, they are in standard frames and meet NEMA torque and current limitations. They are sold "off the shelf." All of them, in fractional- or integral-horsepower sizes, are expected to be operated under normal environmental conditions as described in Chapter 12. They are guaranteed for one year. Neither the manufacturer nor the seller has any idea as to what the buyer will do with the motor once he has taken it home or to his shop. He may hose it down, overload it severely, and burn it out in six months. The manufacturer rarely has any means of proving misuse, and as a matter of good-will, he usually replaces it free of charge.

The definite-purpose motor, as the name implies, is built for a specific application. This does not cover a motor built for one user, as, for instance, the application described in Chapter 14. A definite-purpose motor is designed *for one specific industry*.

The specific definitions of definite-purpose and special-purpose motors as given by NEMA are the following:

MG1-1.08 definite-purpose motor. A definite-purpose motor is any motor designed, listed, and offered in standard ratings with standard operating characteristics or mechanical construction for use under service conditions other than usual or for use on a particular type of application.

MG1-1.09 special-purpose motor. A special-purpose motor is a motor with special operating characteristics or special mechanical construction, or both, designed for a particular application and not falling within the definition of a general-purpose or definite-purpose motor.

An oil burner motor is an example of a standardized definite-purpose motor having important dimensions, ratings, etc., set by NEMA in cooperation with the trade association manufacturing oil burners.

15-1   The Oil Burner Motor

This motor is an example of a standardized definite-purpose motor. By general agreement, motors built for domestic oil burners are of the split-phase type, built in ratings of $\frac{1}{12}$, $\frac{1}{8}$, and $\frac{1}{6}$ hp, operating at 1725 rpm on 60 cycles and designed for 115 and 230 v (see Figure 15-1).

They are expected to operate with the usual NEMA limits on starting current, with the normal values of maximum torque for their horsepower rating. Rotation is clockwise, facing the end opposite the drive. Manually reset thermal protectors are to be provided and the nameplate should so indicate.

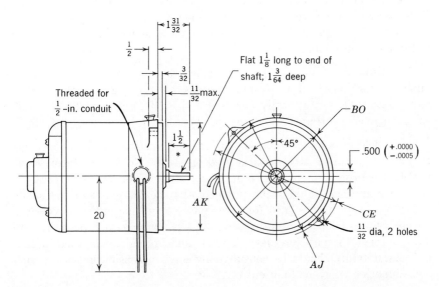

Figure 15-1. Dimensions for face-mounted motors for oil burners, Types M and N. Note the rabbet-fit for mounting with a diameter AK. Two lugs extend beyond this for holding the motor to the pump, on a diameter of AJ. All dimensions in inches. If the shaft-extension length of the motor is not suitable for the application, it is recommended that deviations from this length be $\frac{1}{4}$-in. increments. (From NEMA MG1-18.289.)

Mechanically they should be totally enclosed or, if open, should be of guarded construction. Temperature rises (class A insulation) are set at 65°C for enclosed and 60°C for guarded motors, with the rises measured by resistance increase. Nominally these would be classed as 55° and 50°C respectively, as explained in Chapter 14.

Table 15-1   Dimensions and Nomenclature

AK (in.)	AJ (in.)	CE, max (in.)	BD, max (in.)	Suffix Letter After Frame Number
$5\frac{1}{2}$	$6\frac{3}{4}$	$7\frac{3}{4}$	$6\frac{1}{4}$	M
$6\frac{3}{8}$	$7\frac{1}{4}$	$8\frac{1}{4}$	7	N

15-2   Other Types: Pump Motors

It is obviously impossible to present complete mechanical and electrical details of all of the definite-purpose motors standardized by NEMA. Outstanding characteristics of the common types used as pump drives are shown in Table 15-2. In addition to those shown in the table, the following are available.

Coolant-Pump Motors. This motor is classified as a close-coupled pump motor in which the pump mounts directly against the C-face motor head. They are totally enclosed and built for either 3600 or 1800 rpm. An open ventilated motor of this type is shown in Figure 12-3.

Canned Motors for Pumps for Nuclear Applications. In this construction a thin, stainless-steel tube is used on the inside of the stator, completely isolating it from the rotor. Pump and motor are in a pressure container and the liquid being pumped fills the air gap and surrounds the rotor. They are designed for three-phase operation only, for voltages up as high as 6600. The entire standards pertaining to these units are quite foreign to more ordinary motors, and specialists in their design and manufacture should be consulted by anyone considering using such a combination (until he finds out the price).

Gasoline-Dispensing Pump Motors. Here the chief problem is the hazard from explosive vapors and they are tested and listed by Underwriter's Laboratories. Capacitor-start, repulsion-start induction, polyphase, and direct current are all approved. One-half hp at approximately 1725 rpm is standard.

Table 15-2 Tabulation of Important Characteristics of Some Definite-Purpose Motors

Characteristics	Oil Burner	Cellar Drainer Sump Pump	Jet Pump	Submersible Pump
Construction	Enclosed or guarded rabbet face mounting with two lugs.	Dripproof. Vertical operation. Bottom head designed for mounting on vertical support pipe.	Dripproof, usually with cover when vertical. Bottom (shaft) end with NEMA Std C-flange; see Figure 15-3.	Special enclosure for operating under water. Rabbet fit with studs for mounting to pump.
Electrical type	Split-phase, 115/230 v	Split-phase, 115/230 v	Split-phase; capacitor-start or repulsion-start. Three-phase or direct current. Volts std.	Split-phase; capacitor-start, polyphase. Volts std.
Horsepower and syn. speeds	$\frac{1}{12}, \frac{1}{8}, \frac{1}{6}$ 1,800 rpm	$\frac{1}{3}$ 1,800 rpm	$\frac{1}{3}$ to 1 with higher ratings common, 2,600 rpm.	4-in. well casing (3,600), single-phase. Polyphase, $\frac{1}{4}$ to 3 hp $\frac{1}{4}$ to 5 hp. 6-in. well casing (3,600), 3 to $7\frac{1}{2}$ hp; 3 to 20 hp.
Starting current	NEMA Standard	Not specified	NEMA Standard	NEMA Standard
Torques	Std. max. torque; no std. starting torque	Max. not less than 32 oz-ft. Start 20 oz-ft or over.	Std. max. torque. Starting torque not specified.	Std. max. torque; starting torque not specified.
Service factor	1.0 usual	1.0 usual	No longer specified but unusually high.	Not specified but unusually high.
Thermal protection	Manual reset	Automatic reset	Single-phase, automatic reset.	No std.
Shaft	Flatted	Flatted	Carbon-steel keyed or stainless-steel threaded.	Slotted or splined.
Rotation	Clockwise, facing end opposite shaft.	Clockwise, facing end opposite shaft.	Clockwise, facing end opposite shaft.	Clockwise, facing end opposite shaft.
Bearings	Usually sleeve	Sleeve or ball but suitable for vertical use.	Ball bearings; vertical or horizontal.	Ball bearings; vertical operation.

Figure 15-2. A totally enclosed fan-cooled gasoline pump motor. (Courtesy of Westinghouse Electric Corporation.)

Unusual mechanical features include a built-in switch with an external operating lever, and a D dimension (height of shaft above base) which is otherwise unique in the industry. It is $3\frac{13}{16}$ in. when mounted on a rigid base. Figure 15-2 shows a typical construction.

15-3   Pumps and the C-Face Mounting

Table 15-3 shows dimensions pertaining to the standardized C-face end head. Note the detail in which all of the important dimensions are indicated for both fractional- and integral-horsepower motors. (Unfortunately, some of these dimensions are now undergoing changes for the new rerated T frames.)

Although this head has been mentioned in connection with pump applications, it is undoubtedly the most commonly used of the NEMA standardized mountings, because manufacturers of other devices directly connected to the motor have found it useful. Two examples are magnetic brakes and gear drives.

Perhaps a bit should be said at this point about some pump characteristics with which the reader may not be familiar. In lifting water from a well, for instance, the C-face mounting is usually used with a "shallow well" or a jet-pump motor. A simple pump is only capable of lifting water from a depth of about 20 ft. This is because it is really atmospheric pressure on the water surface which forces water up the pipe to the pump when the running pump first creates the lifting vacuum in the pipe. This is a "shallow-well" pump.

Three means are available to lift water (or other liquids) from a greater depth (see Figure 15-3).

1. *The jet pump.* This installation requires two pipes in the well casing. One pumps water down the well, terminating in a nozzle directed upward in the bottom of the main supply pipe. The velocity of this jet "induces" a flow of water up the pipe. Hence a fraction of the water obtained at the well head is always being pumped down again to continue the jet action.

This is a very common and popular type of pump, effective up to several hundred feet, but losing in efficiency as the depth increases.

2. *The comparatively new submersible motor and pump combination.* Here a pipe line and the electric conductors are connected to the pump and motor respectively, and the assembly is lowered to the bottom of the well. Theoretically there is no limit to the height to which the pump can *push up* water, excepting the pump design and available horsepower. The starting relay, pressure switch, and capacitors, if used, are naturally located in a control box at the top of the well.

3. *Pump at well bottom, motor at ground level,* In these cases one of two types of pumps may be used, rotating or reciprocating. In the former case a long shaft is used from the motor to the pump. Shaft "whip" may have to be minimized by intermediate bearings in the well casing.

In case a reciprocating-type pump is used, a long rod connects the pump with a motor-driven reciprocating mechanism at the well head. This is most common in the oil industry (see Chapter 5).

15-4   Definite-Purpose Motors for Fan Drives

These fall into two general classes: motors for belt-driven or for direct-connected fans or blowers. Such motors represent large production, although in many cases it is a "captive" business. That is, the manufacturer makes both the motor and the fan or blower and may have

Table 15-3

FACE-MOUNTED MOTORS A-C, type C
for integral-horsepower motors

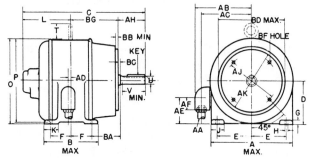

Frame Number	AH	AJ	AK	BB Min
42C‡	1⁵⁄₁₆‡	3¾	3	⁵⁄₃₂*
48C‡	1¹¹⁄₁₆‡	3¾	3	⁵⁄₃₂*
56C	2¹⁄₁₆‡	5⅞	4½	⁵⁄₃₂*
66C	2⁷⁄₁₆‡	5⅞	4½	⁵⁄₃₂*
182C	2⅛	5⅞	4½	⁵⁄₃₂*
184C	2⅛	5⅞	4½	⁵⁄₃₂*
213C	2¾	7¼	8½	¼
215C	2¾	7¼	8½	¼
254UC	3½	7¼	8½	¼
256UC	3½	7¼	8½	¼
284UC	4⅝	9	10½	¼
286UC	4⅝	9	10½	¼
324UC	5⅜	11	12½	¼
324SC	3	11	12½	¼
326UC	5⅜	11	12½	¼
326SC	3	11	12½	¼
364UC	6⅛	11	12½	¼
364USC	3½	11	12½	¼
365UC	6⅛	11	12½	¼
365USC	3½	11	12½	¼
404UC	6⅞	11	12½	¼
404USC	4	11	12½	¼
405UC	6⅞	11	12½	¼
405USC	4	11	12½	¼
444UC	8⅜	14	16	¼
444USC	4	14	16	¼
445UC	8⅜	14	16	¼
445USC	4	14	16	¼
504UC	8⅜	14½	16½	¼
505C	8⅜	14½	16½	¼
504SC	4	14½	16½	¼
505SC	4	14½	16½	¼

AH dimensions in inches.

* These BB dimensions are miximum dimensions.
† These BD dimensions are nominal dimensions.
‡ Suggested standard for future design.

TOLERANCES
AK Dimension—
42C to 286UC frames, incl., +0.000, −0.003 inch.
324UC to 505SC frames, incl., +0.000, −0.005 inch.
Face Runout—
42C to 286UC frames, incl., 0.004-inch indicator reading.
324UC to 505SC frames, incl., 0.007-inch indicator reading.

Note: for dimensions of direct-current motors having Type C face mountings, the motor supplier should be consulted.

Table 15-3 (Continued)

for motors where the AJ dimension is greater than the AK dimension

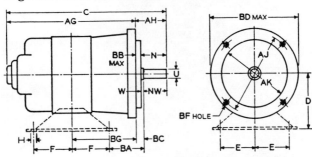

BC	BD Max	BG	BF Hole Number	BF Hole Tap Size	Bolt Penetration Allowance
3/16	5†	2 23/32	4	1/4–20	......
3/16	5 5/8†	3 11/16	4	1/4–20	......
3/16	6 1/2†	4 1/16	4	3/8–16	......
3/16	6 1/2†	5 7/16	4	3/8–16	......
—1/8	6 1/2†	....	4	3/8–16	9/16
—1/8	6 1/2†	....	4	3/8–16	9/16
—1/4	9	....	4	1/2–13	3/4
—1/4	9	....	4	1/2–13	3/4
—1/4	10	....	4	1/2–13	3/4
—1/4	10	....	4	1/2–13	3/4
—1/4	11 1/4	....	4	1/2–13	3/4
—1/4	11 1/4	....	4	1/2–13	3/4
—1/4	14	....	4	5/8–11	15/16
—1/4	14	....	4	5/8–11	15/16
—1/4	14	....	4	5/8–11	15/16
—1/4	14	....	4	5/8–11	15/16
—1/4	14	....	8	5/8–11	15/16
—1/4	14	....	8	5/8–11	15/16
—1/4	14	....	8	5/8–11	15/16
—1/4	14	....	8	5/8–11	15/16
—1/4	15 1/2	....	8	5/8–11	15/16
—1/4	15 1/2	....	8	5/8–11	15/16
—1/4	15 1/2	....	8	5/8–11	15/16
—1/4	15 1/2	....	8	5/8–11	15/16
—1/4	18	....	8	5/8–11	15/16
—1/4	18	....	8	5/8–11	15/16
—1/4	18	....	8	5/8–11	15/16
—1/4	18	....	8	5/8–11	15/16
—1/4	18	....	4	5/8–11	15/16
—1/4	18	....	4	5/8–11	15/16
—1/4	18	....	4	5/8–11	15/16
—1/4	18	....	4	5/8–11	15/16

Permissible Eccentricity of Mounting Rabbet—
42C to 286UC frames, incl., 0.004-inch indicator reading.
324UC to 505SC farmes, incl., 0.007-inch indicator reading.

Permissible Shaft Runout—
182C to 286UC frames, incl., 0.002-inch indicator reading.
324UC to 505SC frames, incl., 0.003-inch indicator.

These dimensions may also be applied to vertical motors in which case the frame designation may have the suffix letters CV, UCV or SCV.

Figure 15-3. A jet pump motor, showing the stainless steel shaft and threaded end *C*-face mounting. (Courtesy of the Westinghouse Electric Corp.)

no need to meet any motor standards beyond production of a satisfactory, marketable unit. On the other hand, many fan motors are sold as such to be used with separately purchased fan blades or blowers.

Direct-Connected Fans. The connection of a fan as a load to the motor shaft almost always results in a good draft of air being blown over or through the motor. As long as the air temperature is below that of the motor and/or the windings, this is beneficial. It can mean a reduction in motor size for a given horsepower. It will be noted that, except for the shaded-pole types (Table 15-4), total enclosure is recommended. A fan running for any length of time, even in apparently clean air, moves a great deal of dust and foreign matter. It can accumulate in an open motor, clogging ventilating passages otherwise needed for cool operation, and can contaminate the lubrication.

Table 15-4 Tabulation of Important Characteristics of Air-Moving Definite-Purpose Motors

Characteristics	Shaft-mounted fan or blower	Shaft-mounted: fan or blower; shaded-pole types	Motors for belt-driven fans and blowers	Air-conditioning condenser and evaporator fans
Construction	Totally enclosed frames. Foot mounted or by extended through-bolts. Capacitor may mount on front head.	Open or totally enclosed. Foot mounted or by extended through-bolts. Three lugs standardized for these and permanent-split-capacitor types.	Open or dripproof resilient mounting	Open or enclosed air over motor. Usually supplied with resilient rubber rings on hubs.
Electrical	Split-phase; permanent-split-capacitor; polyphase (all types std. volts); Ac shunt or compound.	Shaded-pole, 115 or 230 v	Split-phase; capacitor-start; repulsion-start. All one speed. 115 and 230 v	Single-phase; shaded-pole and permanent-split-capacitor, 115 and 230 v.
Horsepower and speeds	$\frac{1}{20}$, $\frac{1}{12}$, $\frac{1}{8}$, $\frac{1}{6}$, $\frac{1}{4}$, $\frac{1}{3}$, $\frac{1}{2}$, $\frac{3}{4}$. Split-phase, polyphase, and dc. 1,725, 1,140, 850 rpm; Permanent-split-capacitor types lower.	1, 1.25, 1.5, 2, 2.5, 3, 4, 5, 6, 8, 10, 12.5, 16, 20, 25, 30, and 40 mhp also $\frac{1}{20}$ to $\frac{1}{4}$ hp; speeds 1,550, 1,050, 800.	Split-phase and capacitor-start. Two-speeds, 1,725/1,140 rpm. $\frac{1}{6}$, $\frac{1}{4}$, $\frac{1}{3}$ hp, split phase. Others $\frac{1}{3}$, $\frac{1}{2}$, $\frac{3}{4}$.	No standard horsepower or speeds, but 2-, 4-, 6-, or 8-poles all used. Multispeeds common.
Starting current	For single-phase NEMA standard	No standard	NEMA standard	No standard
Torques	Max. torque standard. No standard on starting torque.	Max. torque standard. No standard on starting torque.	Max. torque standard. No standard on starting torque.	No standard
Service factor	No standard. Usually 1.0	No standard. Usually 1.0	NEMA Standards. Usually 1.0	No standard. Usually 1.0
Thermal protection	No standard	No standard	Automatic reset	No standard
Shaft	Standard NEMA for frame size used	Standard in frames covered by NEMA	Standard NEMA for frame sizes used	Standard with flat
Rotation	Either direction	Counterclockwise, facing end opposite drive.	Either direction	No standard
Bearings	Sleeve with axial thrust provisions. Ball if vertical.	Sleeve bearings	Sleeve bearings	Sleeve bearings

Direct-Connected Blowers. It is assumed that the reader is familiar with the radial-type blower wheel, sucking in air at one end and whirling it out into a stationary spiral casing. The latter is usually supplied with a square or rectangular flange to which duct work can be connected.

It has always been a mystery to the author as to why nearly all such blowers are constructed with the air inlet on the side away from the motor. As a result, the driving motor is operating in almost "still air," with no additional cooling benefit by having the air input to the blower drawn over or through the motor surfaces (see Figures 15-4 and 15-5).

No better proof of this shortcoming is needed than to take a given horsepower fan motor and apply it to a blower of the same power requirement and then note that the motor becomes too hot for satisfactory operation.

Figure 15-4. A direct-connected furnace-blower motor with speed change by turning the female receptacle for the male terminals. Note the exposed laminations and the absence of a motor body or shell. (Courtesy of Westinghouse Electric Corporation.)

Figure 15-5. Permanent split capacitor motors used for direct fan drive or for double end blower duty. (Courtesy of A. O. Smith.)

Speed Control. Both the shaded-pole motor and the permanent-split-capacitor types are quite satisfactory for adjustable-speed fan or blower operation. In such cases standards are set up for external blade-type connectors on the motor with a spacing and terminal sequence as agreed upon in the industry.

15-5 Belt-Driven Fans and Blowers

These are commonly used with hot- or cold-air circulating systems and attic ventilators. As in nearly all fan or blower applications, quietness of operation is important. In some cases, depending on the distance between the motor and the fan shafts, the motor benefits in cooling by being in the air stream. This is especially true, of course, in the case of attic ventilators or cool-air circulating systems. Blower motors obtain no such benefit.

It will be noted that only motors with relatively high starting torques are recommended for this service. (Permanent-split-capacitor and shaded-pole motors do not qualify for belt drives.) This is because a tight belt may readily present too great a friction load for the system to start.

Speed change is obtained only by two windings in the motors. In all cases (split-phase or capacitor-start) a centrifugal switch or starting relay must be so constructed that it operates below the 1,140-rpm range. Otherwise it would remain closed and the start winding would be connected in the circuit continuously (with rapid burnout) when the motor was operating on its 1,140-rpm connection. No intermediate speeds are practicable because of the presence of this switch or relay.

15-6   Hermetic Motors

These are really a set of parts defined by NEMA:

A hermetic motor consists of a stator and rotor without shaft, end shields or bearings for installation in refrigeration compressors of the hermetically sealed type.

The dictionary definition of hermetic is belatedly useful.

(1) Relating to Hermes Trismegistis and his lore; (2) magical, alchemical; (3) completely sealed . . . so that no gas or air can escape or enter.

It is safe to assume that the third meaning is applicable here, because the motor parts are built right into the refrigerating system so that they are exposed to the refrigerant, are cooled by it, and have no exposure to the outside air.

At first such motor parts were used only in the fractional-horsepower ratings as found in household refrigerators, etc. Larger standard motors were used for belt-driven compressors. Then as the economics of the sealed-in hermetic motor parts became impressive, refrigerating units requiring up to 125 hp became available.

Heating of the motor windings is not a direct problem. They are always cool because they are in the refrigerant. But on the other hand, the losses become a load on the refrigerating system so that an inefficient set of parts could mean fewer British thermal units absorbed per kilowatt-hour input. This is important in this highly competitive business. A few extra watts loss (and a few extra pennies in price) may determine whether motor manufacturer A makes 50,000 sets of parts or 500,000 sets this year.

With the losses taken care of in the manner described, temperature rise as one of the limits on horsepower rating is no longer a problem. Instead, motors are identified in terms of maximum or breakdown torque and starting current. It is of interest to note the range available (see Table 15-5a and b).

Miscellaneous Standard Items. Electrical types are expected to be split-phase, capacitor-start, capacitor-start and capacitor-run, or permanent-split-capacitor. (Capacitor-run characteristics are obviously useful for improving breakdown torques and reducing losses.)

These single-phase parts are for operation on 230 or 115 v. At the present time, polyphase parts are expected to operate on 208, 220, 440, or 550 v. Thermal protectors, if used, are expected to be tied into the motor end coils.

Standardized diameters of the stator, either laminations or the thin-shelled sometimes used over the stack, are (in inches) 4.792, 5.480, 6.292, 8.777, 10.125, 12.375, 15.562.

In considering the sale of universal motor parts for portable tools,

Table 15-5a   Single-Phase Hermetic Motors*

	1800 rpm			3600 rpm	
Breakdown Torque (oz-ft)	Locked-Rotor Current, Amperes at 115 v		Breakdown Torque (oz-ft)	Locked-Rotor Current, Amperes at 115 v	
10.5	20		5.25	20	
12.5	20		6.25	20	
15	20		7.5	20	
18	20		9.0	20	
21.5	20		10.75	21	
26	21.5		13.0	23	
31	23		15.5	26	
37	28	23†	18.5	29	
44.5	34	23†	22.0	33	
53.5	40		27.0	38	
64.5	48	46†	32.0	43	
77	57	46†	38.5	49	
92.5	68	46†	46.0	56	
Breakdown Torque (lb-ft)	Locked-Rotor Current, Amperes at 230 v		Breakdown Torque (lb-ft)	Locked-Rotor Current, Amperes at 230 v	
7	36		3.5	32	
9	38		4.5	39	
11	44		5.5	46	
14	56		7.0	56	
18	68		9.0	69	
23	85		11.5	85	
29	104		14.5	104	
36	126		18.0	126	
45	155		22.5	154	

* Hermetic motor parts used in refrigerating systems are rated by these characteristics rather than by horsepower. (From NEMA MG1-18.081.)
† Motors having locked-rotor currents within these values usually have lower locked-rotor torques than motors with the same breakdown torque ratings and the higher locked-rotor current values.

Table 15-5b   Polyphase Squirrel-Cage Induction Hermetic Motors*

1800 rpm		3600 rpm	
Breakdown Torque (lb-ft)	Locked-Rotor Current, Amperes at 220 v	Breakdown Torque (lb-ft)	Locked-Rotor Current, Amperes at 220 v
9	24	4.5	24
11	30	5.5	30
14	38	7.0	38
18	48	9.0	48
23	59	11.5	59
29	71	14.5	71
36	85	18.0	85
45	102	22.5	102
56	125	28.0	125
70	153		
88	189		

* Hermetic motor parts used in refrigerating systems are rated by these characteristics rather than by horsepower. (From NEMA MG1-18.081.)

etc., it was pointed out that a great many vital dimensions had to be known by the manufacturing user before the latter could design his equipment. This is equally true for hermetic parts; hence detailed dimensions are available but not included here.

Cleanliness is of great importance in the production of these parts. Not only must all components of the insulation be compatable with the refrigerant used, so that deterioration does not cause flaking off, but the usual lint, dust, chips, etc., found in ordinary motors could choke up the expansion nozzle needed in the refrigerating cycle and ruin the operation. Hence cleanliness standards are established in great detail.

15-7   Home-Laundry Equipment Motors

These are intended for use on domestic washing machines, dryers, or a combination of both.

These motors represent some of the highest volume business in definite-purpose motors, and thus competition is high among a fairly small portion of the electric motor industry (see Chapter 8).

Motor Types. Single-phase only, split-phase or capacitor-start, for voltages of 115 and 230 v. They are built for single speed, approximately 1725 rpm, or for double speed for 1725/1140 rpm.

Horsepower and Torques. Horsepower rating are standardized at $\frac{1}{6}$, $\frac{1}{4}$, $\frac{1}{3}$, and $\frac{1}{2}$ hp. In each case the minimum value of breakdown torque is fixed as is also the minimum value of starting or locked rotor torque.

Starting Current. This motor type is allowed to take as high as 50 amp at starting. It is also allowed a 50°C nominal rise in temperature with class A insulation and 70°C with class B (60 and 80°C respectively if the rises are measured by resistance).

Mechanical Construction. These motors are available in NEMA frames 56 or 48, with the trend being toward the latter size. They are open-ventilated with sleeve bearings and are expected to be supplied with resilient rings on their hubs having an outside diameter and spacing as shown in Figure 15-6. An alternate mounting arrangement makes use of extended through-bolts, with distances between them of 4.08/3.98 in. on the square.

Because of the high degree of production tooling and automation provided for their high demand, these motors probably present the greatest bargain in the motor industry.

15-8   Other Definite-Purpose Motors

The foregoing items represent the most common types of NEMA definite-purpose motors (with the exception of the canned motor for nuclear

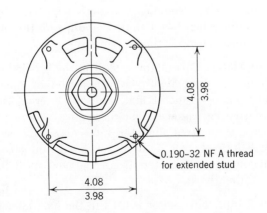

Figure 15-6. Dimension for motors for home laundy equipment, Type *L*.

pump applications, which was included as an example of a unique design).

Other motors with which the reader is likely to have less contact are ac and dc elevator motors, M.G. sets for the latter, crane motors, and motor parts for wood-working machinery. In addition, some standards are set up for instrument and gear motors. These will be covered in later chapters.

15-9 Special-Purpose Motors

These involve modified near-standard motors in which some electrical and/or mechanical modifications are made. As they are not all covered by any standards, the possibilities are almost limitless.

Four examples of common special-purpose motors are the pancake, the vertical-hollow-shaft motor, the lint-free motor, and the sanitary motor. Of these only the vertical-hollow-shaft motor is covered dimensionally by NEMA. They all represent mechanical modifications.

Pancake Motors. They represent designs, as the name implies, which are very flat. The end coils are compressed as tightly as possible against the stack of laminations, no room is provided for a full-diameter fan, and the bearings are "inboard." The result is a flat motor, usually face-mounted, which has found considerable usage in the machine tool industry.

The Vertical-Hollow-Shaft Motor. This motor is used at the well head with a turbine pump which has a shaft that slips up through the hollow shaft of the motor and is coupled thereto. The entire weight of the pump mechanism and the column of fluid is supported by the motor, with a turbine pump which has a shaft that slips up through the hollow on the flanged mounting used as the bottom head.

The Lint-Free Motor. This motor is used in the textile industry which is cursed with lint floating about. Over a period of time enough lint is pulled through an open motor and sticks to internal rough surfaces so as to clog ventilating passages. The mechanical modification required in an open motor is to streamline the internal air passages such that lint goes through it freely. Or if totally enclosed fan-cooled motors are used, as is common, again the external fan and cowl must be so designed as to prevent clogging the air passages.

The Sanitary Motor. This motor is used in the food industry and must be so constructed that it can be hosed down. But more than that, the external surfaces must be so nicely rounded that no corners or pockets are available in which food stuff (or the components thereof) can lodge and decay.

15-10 Other Mechanical Modifications

Mechanical changes in motors can involve new end heads, special mounting arrangements, and special methods or degrees of enclosure. Years ago when the motor industry made a fair profit,[1] it was not unusual for a motor company to welcome such business even though quantities might be small. Today, with standardizing on frames and high costs of even temporary low production tooling, the contrast between costs of such special constructions as compared with the costs of standard motors is so great as to discourage the acceptance or fulfillment of such business.

There are two ways to purchase highly modified motors for which only a low production is required. One is to pay the motor company for the patterns or other tooling involved. The other method is to obtain the required motors at reasonable prices because the motor manufacturer is willing to absorb the loss for a highly rated, otherwise good customer. A great deal of the latter business goes on in the motor industry. An OEM account buys thousands of motors from a supplier, but has been forced for one reason or other to continue the manufacture of one of its old products which it would prefer to make obsolete. It is not expedient to do so and, hence, every year or so the motor manufacturer receives an order for a few motors of special design which have not been standard for perhaps 30 years. It takes many years for an old special-purpose motor to disappear from production, for reasons such as given previously.

15-11 Electrical Modifications: Special-Purpose Motors

The modifications usually take the form of requiring engineering and testing time rather than tooling expense. Two good examples would involve a motor built for frequent starting and reversal by plugging, and an intermittent-duty motor for which more than usual horsepower is designed into a smaller frame for short-time operation. At best, these results are obtained by a computer design and no new tooling. At the worst, they may prove too hot by test and require a redesigned ventilating system.

[1] This goes back before the memory of the author.

Two other common electrical specials involve methods of starting large polyphase motors with less than normal starting current.

Part-Winding or Increment Start. Consider a motor with two sets of windings (as is usually the case) connected in parallel for 220 v or in series for 440 v. When used for increment starting such a motor can be used only on 220 v with the two sets of windings in parallel.

But only one set is connected to the line to get the motor started. The starting current is a little over half normal. Then an instant later the second winding is connected and the motor operates normally from that point on.

All of this is done automatically with two sets of magnetic contactors

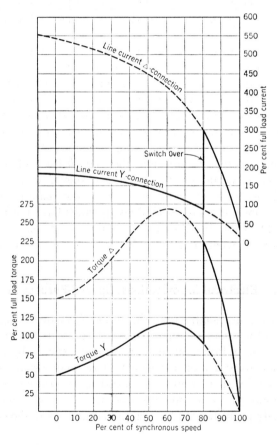

Figure 15-7. Y-Δ starting. The solid lines show the current on switching over from Y to Δ, when the motor has reached 80 percent of synchronous speed.

and an adjustable timer between. The windings should be Y-connected because the controls are thereby simpler and less costly than for delta-connections.

Y-Start–Delta Run. If a motor is normally designed for delta connection, but its windings are connected in Y for starting, the increased impedance of the winding reduces the starting current. Sometime during the starting period, a simple contactor or manual switch can be used to change over from Y to delta. Resulting characteristics are shown in Figure 15-7.

Chapter 16

Servo and Instrument Motors

A servomotor is essentially a small, two-phase motor, designed with low inertia for rapid response. It has small-diameter higher-resistance rotors, resulting in quick response to start, stop, and reversal. The high resistance is essential to a near linear speed-torque curve for accurate control.

It is well known that rotor resistance does not affect the value of maximum torque, but only the speed at which it occurs. Curve 3 in Figure 16-1 shows a suitable rotor resistance for servo operation because it results in a nearly linear speed-torque curve.

Such a motor is wound with its two windings at right angles. One winding is excited by an adjustable voltage control. The windings are usually designed with equal impedances so that power inputs at maximum fixed phase input and at maximum control phase input are in balance. The purpose, of course, is to obtain adjustable speeds and torques. Ideal speed-torque curves are shown in Figure 16-2. Performance curves for a typical 400-cycle servomotor are displayed in Figure 16-3.

Ideally the torque at any speed is directly proportional to the control winding voltage. Actually, this condition exists only at zero speed (starting torque). The error is shown in contrasting the slopes of the curves of Figures 16-2 and 16-3.

16-1 Stalled Torque

The results of the unbalanced input wattage on the starting or stalled torque may be calculated from

$$\frac{T_{su}}{T_{sb}} = \frac{W_{cu}W_{fu}}{W_{cb}W_{fb}}, \tag{16-1}$$

where b = balanced input, c = control
  u = unbalanced input, f = fixed.

266

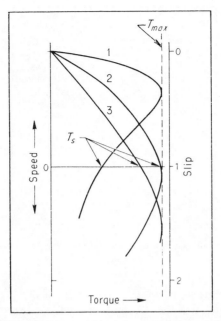

Figure 16-1. Effect of rotor resistance on speed-torque characteristics. Motor 1 has a normal rotor resistance; 2 has medium resistance; and 3 has high resistance. (Reference Issue, *Electric Motors,* Machine Design 1965)

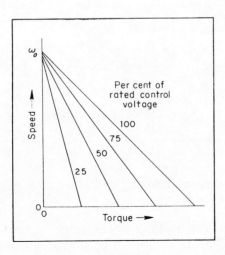

Figure 16-2. Family of speed-torque curves for various control-winding excitations.

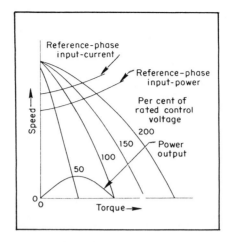

Figure 16-3. Performance curves for a typical size 11, 400-cps servomotor.

Hence with a total input of 12 w and with a 50% unbalanced input, the resulting torque will be $(3 \times 9)/(6 \times 6)$ or 0.86 of the balanced stalled torque. In general, Eq. 16-1 becomes incorrect with a high degree of unbalance, and direct measurements may be necessary. In considering unbalanced operation, these factors must be given consideration: increased heating under conditions of fixed phase excitation, loss of torque, increased response resulting from a higher torque per control watt.

16-2   Dynamic Constants

The time required for a motor to accelerate from rest to 63% of its no-load speed is

$$t_n = \frac{JN_0}{T_s} \times 1.485 \times 10^{-6}, \qquad (16\text{-}2)$$

where t_n = time, sec,
  J = rotor inertia, g-cm^2,
  N_0 = no-load speed, rpm,
  T_s = stalled torque, oz-in.

Reversing Time. This is defined as the time required for the motor to reverse from 63% of its no-load speed in one direction to 63% of its no-load speed in the opposite direction. Approximately

$$t_r = 1.7t_n. \tag{16-3}$$

16-3 Damping

Internal damping of the motor decreases as the control phase voltage is decreased (see Figure 16-3). In the slow-speed region the slope of the speed-torque curve is such that damping is not sufficient to maintain stability. To cure, this, various means are employed in motor assemblies to provide stabilizing voltages.

Damping Tachometers. A field and drag cup are used for this purpose, although they have certain disadvantages. Reaction of the field and drag cup load the motor in proportion to the speed. This limits the response speed of the system. Also the length of the servo and its cost are increased. The excitation phase of the tachometer requires power which adds to the heating.

Viscous Damping. This is provided integral with the motor by means of a drag cup mounted to the motor shaft rotating in the field of a permanent magnet. Interaction between the field and the drag-cup eddy currents provides damping proportional to the speed. It does not contribute additional heat to the unit.

Inertially Damping. To make use of this type of damping requires a servomotor and a flywheel viscously coupled to the motor shaft. A drag cup is mounted on the motor shaft. Under steady-state conditions motor shaft and damper are operating at the same speed; hence, damping is effective only during a change in speed.

As no excitation is required, power consumption and added heating are not problems. The disadvantage of this type of damping lies in its increased length owing to having two independent rotating elements.

16-4 Power Servomotors

Figure 16-4 shows typical performance for a 25-w and a 200-w 60-cycle servomotor. The inherent damping of these motors is decreased as their rating increases when the power input becomes significant. The motors

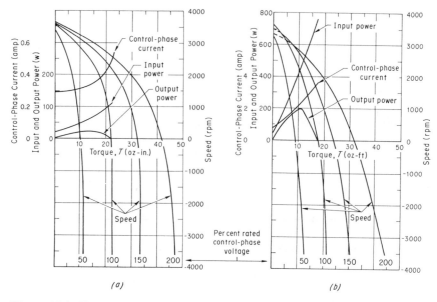

Figure 16-4. Typical performance curves for a 25- and 200 w-servomotor, both 60 cycles. (From *Electric Motors,* Reference Issue *Machine Design,* 1965).

are, therefore, designed for better efficiency at the expense of speed-torque linearity. Integrally mounted blowers are provided with the larger servomotors to maintain safe operating temperatures.

Typical performance characteristics of a number of 60- and 400-cycle servomotors are tabulated in Table 16-1.

16-5   Instrument Motors

A common application for this type of motor is a drive for recording charts. Here a synchronous motor is most commonly used. Recorder pens are usually driven by servomotors. They may be nearly standard two-phase motors with precision low-inertia rotors, as previously described. Response speed is the critical item in determining the motor type.

Larger sizes of motors under this classification are used for combustion control, positioning valves, and dampers; voltage regulators which may change taps on a variable-voltage transformer or valve-positioning motors for opening and closing valves.

Table 16-1   Typical Performance Characteristics of Servomotors

Nominal OD (in.)	Input Frequency (cps)	Rotor Inertia (WK^2)	No-Load Speed (rpm)	Input Power (w)	Stall Torque (oz-in.)
0.5	400	0.175 g-cm²	9,500	3.4	0.1
0.8	400	0.49 g-cm²	6,200	6.4	0.2
1.1	60	0.76 g-cm²	3,000	7.0	0.7
1.1	400	1.07 g-cm²	6,200	7.0	0.6
1.5	400	3.3 g-cm²	5,300	12.2	0.3
1.8	60	4.0 g-cm²	3,400	16.6	3.5
1.8	400	4.0 g-cm²	9,800	32	2.8
2	60	0.14 oz-in.²	3,500	66	12.5
2	400	0.1 oz-in.²	7,500	90	5.3
$3\frac{7}{16}$	60	0.58 oz-in.²	3,300	128	20
$3\frac{7}{16}$	400	0.58 oz-in.²	7,900	180	20
$4\frac{1}{8}$	60	1.3 oz-in.²	3,500	280	55
$4\frac{1}{8}$	400	1.3 oz-in.²	7,500	290	25
$4\frac{5}{8}$	60	2.2 oz-in.²	3,500	530	90
$4\frac{5}{8}$	400	2.2 oz-in.²	7,600	420	42
$4\frac{9}{16}$	60	6.6 oz-in.²	3,500	890	180
$4\frac{9}{16}$	400	6.6 oz-in.²	7,700	860	95
$6\frac{3}{8}$	60	22 oz-in.²	3,500	2,200	540
$6\frac{3}{8}$	400	22 oz-in.²	7,700	1,660	190
7	60	57 oz-in.²	1,700	3,400	1,700
7	400	57 oz-in.²	7,800	2,400	320

From Reference Issue on Electric Motors, Machine Design 1965.

16-6   Motor Types

Instrument motors are available in all electrical types for both alternating current and direct current, and in the absence of any exact definition, they range over all sizes from 1/1,000 hp clock drives up to ¾ hp.

Chapter 17

Special Problems of Large, Single-Phase Motors, Farm, Oil Field, and Similar Applications

17-1 The Need for Large, Single-Phase Motors

For many years the biggest demand for large, single-phase motors was from the oil-pumping industry. They operated in remote areas where the scarcity of other industrial demands did not justify the supplying of three-phase power. Next in this group was the farmer. However, increased mechanization on the farm has now pushed the connected single-phase motor load on the farm to the top position.

Movement to the suburbs in less intensely developed areas has also built up the need for integral-horsepower single-phase motors for filling stations, small shopping centers, and small industry.

17-2 The Problem of Starting Current

It might appear that these needs would impose no new problems. The capacitor-start motor, built in fractional-horsepower sizes, must just be designed with more torque and more material (as is done with the polyphase motor), and we end up with a 10-, 15-, or even a 20-hp single-phase motor.

It is not that simple. Consider the starting current. In Chapter 13 we saw that NEMA recommends limits on the allowable starting current per horsepower. Depending upon the starting torque required, it will be noted that design L will permit 520 amp at starting and design M will permit 400 amp for a 20-hp motor.

A three-phase motor of the same rating is allowed 290 amp., which is all that is required to produce high starting torque.

The *full-load* current of a 20-hp single-phase motor will be 85 to 90 amp. With this as a normal load, when the farmer is operating such a motor, think of the voltage drop in the lines (and the dip in the neighbors' lights!) when he starts this motor and draws, say, 500 amp.

The result is that, although NEMA may suggest such starting currents as given in Table 13-4, many power companies and the Rural Electrification Authority (REA) will not permit such large motors to be connected to their lines. Or, if allowed to operate, special means must be provided for easing them only the line without drawing, say, more than 300 amp at 230 v. (Requirements vary in different localities.)

17-3   "Soft Start" for Large Motors

To meet the conditions described in the foregoing, the motor manufacturer can do several things. Conceivably a large transformer could be used with the motor, having variable voltage taps. The motor and load could be started on reduced voltage, and then, as it gains speed, the voltage could be raised by manual control. This would be expensive. A less expensive plan involves reconnection of the motor windings. This will be described in detail.

Consider a special 3450-rpm single-phase motor designed for 230 v with a nominal rating of 15 hp, but capable of carrying loads of 20 hp for several hours.

The main winding of this motor is shown in Figure 17-1. We obtain exactly the same starting and operating conditions if the winding is made up of single sets of coils of large wire, or double sets of coils with wire of half the cross section. In fact, using multiple sets of windings is a frequent practice to reduce the wire size for ease in handling and to provide for double-voltage connections.

Now refer to Figure 17-2. This motor, normally operating with the two main windings in parallel, shows them in series. The impedance is now four times that shown by the parallel connection, and the inrush of starting current is reduced to one-fourth the previous value.

This method, then, depends on using the main windings in series at start, and switching at about two-thirds of final speed to the parallel running condition. A simple centrifugal switch cannot perform all of these functions, but it can trigger magnetic contactors which can handle the heavy currents involved. The schematic diagram is shown in Figure 17-3. See Figures 17-4 and 17-5 for typical performance curves.

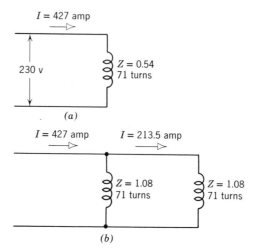

(a)

(b)

Figure 17-1. Considering only the main winding of this large motor, the perform- ance will be exactly the same with one winding or with two, arranged as shown. (a) Main winding, all turns in series. (b) Alternate scheme with smaller wire and two halves of the winding in parallel.

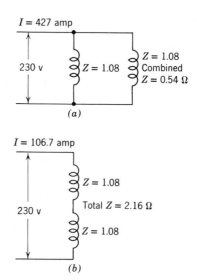

(a)

(b)

Figure 17-2. The parallel windings in the normal running connection display four times the impedance when connected in series. (a) Normal running connection, main only. (b) Special series connection of mains at starting.

274

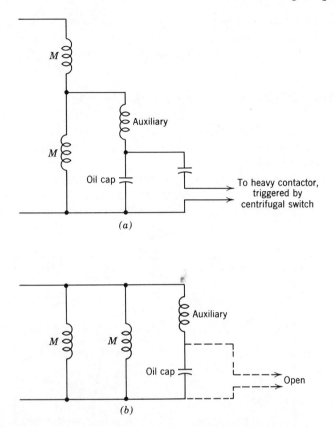

Figure 17-3. Schematic arrangements of "soft-start" capacitor motor. Heavy contactors are required for opening the electrolytic capacitor circuit and for switching the main windings from series to parallel. The circuit returns to the starting conditions when reduced speed reoperates the centrifugal switch. (*a*) Starting connections. (*b*) Running connections.

17-4   Low Starting Torque

One serious difficulty occurs in the use of such motors. If a large starting torque is required, neither the external transformer nor the series parallel winding proves very useful. In each case the applied voltage is reduced, and the starting torque reduces as the square of the voltage. Note in the latter case that the voltage on the auxiliary windings remains the same, starting or running. But the volts per turn on the main winding are one-half the running value, and the starting torque is approximately one-quarter of the full voltage torque.

This scheme can be used for easing a motor onto the line if it can be loaded after it is running or for such loads as fans (used for crop drying), because their starting torque requirements are very low.

17-5   The Repulsion Motor and Modifications

In dealing with various kinds of single-phase motors in Chapter 7, we ignored this type of motor, partly because the dc motor and armature were to be explained first.

The *repulsion motor* is one with a wound armature and commutator similar to the dc construction. But the brushes are short-circuited to each other, providing a path through which the armature current can

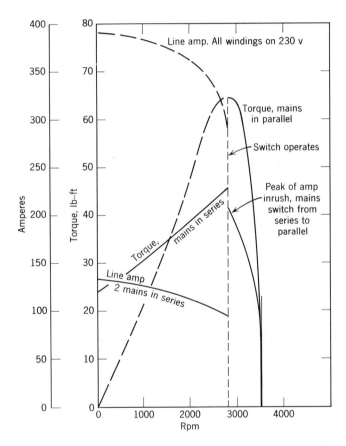

Figure 17-4. Performance "soft-start" motor shows changes during switching cycle.

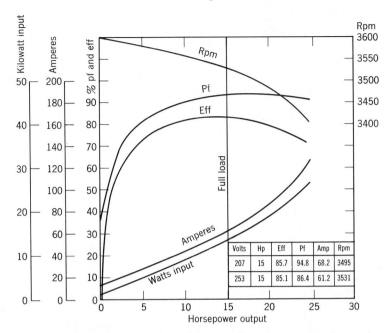

Figure 17-5. Running performance of 15-hp, single-phase, "soft start" motor also shows tabulated data on operation at other than 230 v.

circulate. The stator or field winding only need be one distributed winding similar in design and construction to the main winding of a capacitor-start motor.

A schematic arrangement of these windings is shown in Figure 17-6. The speed-versus-horsepower curve displays speeds above synchronism at light loads and a comparatively low speed under heavy loads. The motor as such is seldom used, but its interesting quality is its very high starting torque per ampere. Under certain conditions it may display twice as much torque per ampere as a capacitor-start motor. Hence it is used mostly as a starting method (see Figure 17-7).

Both starting and running characteristics are influenced by the angle of brush shift. As in the case of the wave-wound dc armature, only two brushes are necessary, even if four- or six-pole stators are used.

Rotation can be reversed by shifting the brushes to an equal angle on the other side of the neutral. If this is impractical on certain applications, two stator windings can be used, each distributed at an angle from the brushes, so that by using one or the other winding the same relative effect is brought about as though the brushes had been shifted mechanically.

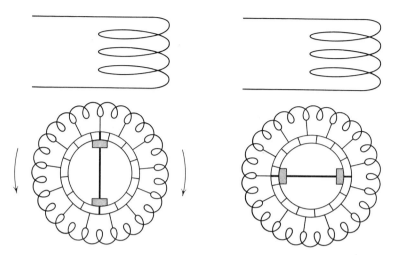

Figure 17-6. Schematic arrangement of the stator and armature windings of a repulsion or repulsion-induction motor. The brushes would normally beat on the outside of the commutator. A compromise in brush position is required to produce useful current in the armature, hence torque. (*a*) With the brushes in this position, voltage induced in the armature winding causes maximum current to flow, but the torques produced by these currents and the main field flux cancel on the opposite sides of the armature. (*b*) With the brushes in this position current flow through the armature and brush circuit would react properly with the field to produce torque, but at standstill no voltage is induced in the armature, hence no current flows.

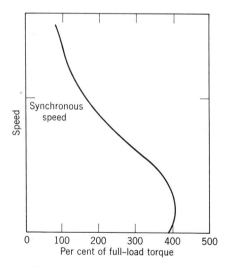

Figure 17-7. Speed-torque curve of a repulsion motor.

17-6   The Repulsion-Start Induction Motor

In this case the motor operates as a repulsion motor up to about two-thirds of final speed. Mechanically and electrically it is identical to the motor just described, except for a centrifugal device which short-circuits all of the commutator bars at the selected speed (see Figure 17-8 and 17-9). Then the wound armature, with its winding short-circuited, acts like a squirrel-cage rotor. It continues its acceleration and operates finally with a slip speed below synchronism. The speed-torque curve is shown in Figure 17-10.

17-7   The Repulsion-Induction Motor

This represents a modification of the above mentioned motor, in which the armature has a conventional squirrel cage below the usual slots

Figure 17-8. Parts of a repulsion-start, induction-run motor showing the centrifugally operated commutator short-circuiter. (Courtesy of Howell Electric Motors.)

Figure 17-9. Cutaway view of the repulsion-start, induction-run motor. (Courtesy of Howell Electric Motors.)

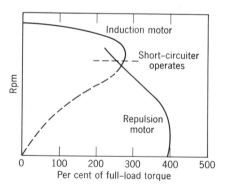

Figure 17-10. As a starting means for a single-phase motor, the torque characteristics are unusually desirable.

for the armature winding. A commutator and shifted brushes are used as before, but there is *no short-circuiter*. It always runs as both a repulsion and induction motor. Starting and accelerating torques are high. At light loads, above synchronous speed, the torque is obtained from the armature winding, whereas the cage acts as a drag to keep down excessive speed. Heavy loads reduce the speed, and below synchronism both the winding and the cage provide torque. The combined effect is a speed change with load somewhat between the sharp slope of the repulsion motor and the conventional squirrel cage. The brushes are in the circuit continuously during operation.

17-8   Availability

Two things are wrong with these motors. They display about the same service requirements as a dc motor with commutator and brush wear. Sparking interferes with radio and television reception.

But the worst thing is that few motor manufacturers wish to build the repulsion-induction motor or the repulsion-start induction motor, and those who do claim it is an unprofitable line.

Before the development of the electrolytic capacitor in the 1930s, almost all motor manufacturers built repulsion start motors, from about $\frac{1}{3}$ hp on up. This was the only means of obtaining high starting torque above the split-phase range. Wound armatures, commutators, brush rigging, and centrifugal short-circuiters are expensive. A capacitor-start motor could be built for less. If a motor manufacturer charged much more for, say, a 5-hp repulsion-start induction motor than for a capacitor-start motor, his customers would try hard to make do with the latter. The business dwindled and, in the meantime, several NEMA rerating programs came about. Retooling for the new designs appeared too expensive for the volume of business and the small profit available, so one by one manufacturers dropped this type of motor from their line.

As previously indicated, the starting torque is high. For instance, 400 to 425% of full load might be obtained with only 90 amp starting current on 230 v for a 5-hp motor. A capacitor motor could hardly do half that well.

The farm market requires high-starting-torque motors with small inrush current for hammer mills, feed grinders, silo unloaders, barn cleaners, and so on. These repulsion-type motors are excellent for this service, but, as mentioned previously, the general level of the market price discourages their manufacture.

17-9   Static-Phase Converters

It is possible to start and operate a three-phase motor from a single-phase supply. To do so requires a *static-phase converter,* usually consisting of starting and running capacitors and a voltage-type relay for disconnecting the electrolytic starting capacitors from the circuit (see Figure 17-11). The voltage relay, in contrast to the current relay for similar duty, has many turns on its actuating coil and is connected *across* the auxiliary winding. The voltage rises on this winding with speed increase (approaching resonance), and the contacts open the auxiliary winding circuit.

In short, the external-phase converter usually contains the same auxiliary equipment found in a single-phase motor.

Their use is largely a matter of economics. Does a static-phase converter plus a polyphase motor cost less than a single-phase motor? If so, the combination may be justified if certain limitations on performance are not serious.

Unfortunately, static-phase converters represent a controversial subject. Users and prospective users have frequently been led to believe that a 5-hp three-phase motor, for instance, will display the same torques and operating characteristics with a converter on single phase as when a normal three-phase supply is used. Some of the early producers of this equipment made such claims, which are unfortunately not true.

Theoretical analyses and tests show the following results on a 5-hp motor:

It appears that the proper rating of a standard three-phase motor with a phase converter, operated on a single-phase supply of the same voltage is not more than:

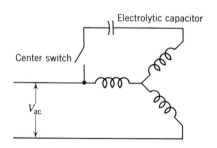

Figure 17-11. A simple form of connections for operating a three-phase motor from a single-phase source.

a. 60% of the three-phase rating if the standard percent locked-rotor torque is required.

b. 60% of the three-phase rating if the standard percent breakdown torque is required.

c. 60 to 70% of the three-phase rating if not more than 10% higher temperature rise (total power loss) is to be tolerated.[1]

In 'short, the 5-hp three-phase motor becomes an equivalent 3-hp three-phase motor.

Such results are not surprising. The designer of electric motors knows that, to obtain satisfactory performance with minimum material, the comparative number of turns in the main and the auxiliary windings must be fixed rather closely and coordinated with the size of the capacitor. Windings of any three-phase motor are identical in the three phases and need not work best with the converter available.

17-10   Static Converters with Auxiliary Transformer

Some of the disadvantages just noted can be overcome by the use of a transformer excited from the single-phase supply with its output connected across the pair of leads of the three-phase motor which are used as the auxiliary winding. Taps on this transformer enable the supply voltage to be adjusted for various loads and to bring about approximately equal currents in all three lines.

Naturally this adds to the cost and, again, the choice is a matter of economics.

17-11   Voltage Drop in a Distribution System

The successful operation of large, single-phase motors depends to a great extent upon maintaining standard voltage at the motor terminals. This is especially true under starting conditions when the large inrush current causes terminal voltages to drop. Occasionally motors fail to start, or operate for a short time at a fraction of their full-load speed owing to a lack of accelerating torque, brought about by low voltage. Protective devices are necessary to take the motor off the line to prevent burnouts under these conditions.

The usual power supply involves a high-voltage line to which a step-down transformer is connected for supplying three-wire 115/230-v power

[1] Quoted from R. Habermann, Jr., "Single Phase Operation of a Three Phase Motor with a Simple Static Phase Converter," AIEE, 1954.

to the farm or other isolated load. The householder should not be confused by these three wires; they do not signify a three-phase supply. They represent a simple system of providing 230 v between the two outside wires and 115 v from the center wire to either outside line as shown in Figure 17-12.

It is not our purpose to go into the problems of voltage regulation (the percentage change in voltage from no load to full load) of distribution systems. This is a large field of knowledge in itself. But it is enlightening to consider briefly the voltage drop in the pair of 230-v lines from the transformer to the terminal box of a large, single-phase motor. Understanding this means that the isolated user of a large motor is less likely to complain, "I've got a weak motor."

An example will be considered in which the following assumptions are made: The high-voltage supply of the transformer is maintained about constant; the voltage drop in the transformer will be ignored.

Actually it may run as high as 2%. This means that a transformer with an output at no load of 240 v might drop 2% or 4.8 v when fully loaded.

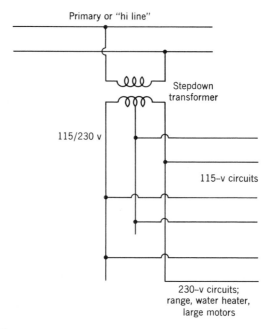

Figure 17-12. Customary arrangement for distributing three-wire 115/230 v power through a transformer from the high-voltage, single-phase supply.

Example   285

It is also assumed that, although other loads are connected to this line being considered, their effect is negligible.

This all means that, in considering the voltage drop in a typical line when a large motor is running or starting, we shall see optimistic results. Actual operation is likely to be worse.

17-12   Resistance and Reactance Drops

We are already aware that the resistance of a line can result in an IR drop. What is not generally realized is that a supply line also has an inductive reactance (IX_L) drop.

A straight length of wire has an alternating field built up around it, which cuts its own cross section, giving itself inductance. Of course, this effect is much less than that shown by the same wire wound into a coil. The effect varies with the distance between the two lines. The current in one wire is naturally the return circuit for the other wire, and the currents flow in opposite directions. The two build up magnetizing forces which oppose each other and, if they are close together, the resulting flux system is smaller than if they were far apart. Values for wire size and spacing are given in the Appendix.

17-13   Selecting Wire Size

The National Electrical Code recommends the current-carrying capacity of various sizes of wire, depending on the insulation used and whether they are in conduit or free air. Only one set of values is given in the Appendix along with the line-reactance table. These values are for slow-burning weatherproof-type cables in free air.

The first step in selecting wire for a motor supply line is to see that it is large enough to meet the code. However, this could still result in a line drop too great to maintain the necessary voltage.

17-14   Example

We shall consider the 15-hp motor described earlier in this chapter, and operating at a 20-hp load. The current is then 84 amp at a power factor of 0.927. This is the cosine of 22°, of which the sine is 0.375. Suppose that the "soft start" connection was not used; then the line current at starting would be 390 amp. This is less than the current in the main winding because of the angle of lead of the auxiliary winding current. The power factor at starting is 0.636. This is the cosine of 50.5° for which the sine is 0.77.

The motor is located 300 ft from the transformer. The lines will be 18 in. apart. The problem is to select the wire size and determine the voltage necessary *at the transformer* to maintain 230 v at the motor terminals, both at starting and running conditions.

Reference to the Appendix shows that No. 6 wire (capacity 100 amp) would be required. (The short-time starting current is ignored in such a choice.) The inductive reactance is 0.1296 Ω/1000 ft of single line. Reference to the wire table shows that the resistance of No. 6 wire is 0.3952 Ω/1000 ft. Therefore, using the full-load current,

$$IX_L = 84 \times \tfrac{600}{1000} \times 0.1296 \text{ or } 6.5 \text{ v,}$$

$$IR = 84 \times \tfrac{600}{1000} \times 0.3952 \text{ or } 20 \text{ v,}$$

$$IZ = \sqrt{(IR)^2 + (IX_L)^2} \text{ or } 21 \text{ v.}$$

Now refer to the phasor diagram of Figure 17-13. Note that the IR drop is in phase with the current and the IX_L drop lags by 90°. The answer is obtained by making the transformer voltage the hypotenuse

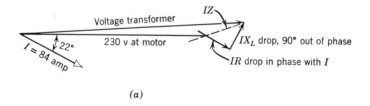

(a)

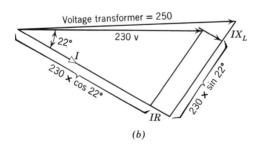

(b)

Figure 17-13. A phasor diagram showing relative positions of voltage drops in the two supply lines. Not to scale. In (b) is shown a method of determining the required transformer voltage when the motor is operating at a 20-hp load.

Example   287

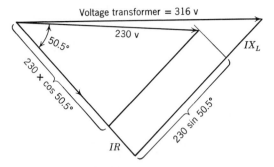

Figure 17-14. Phasor diagram showing transformer voltage required to maintain. 230 v at the motor terminals under starting conditions.

of the triangle made up of fictitious in-phase and out-of-phase components. This shows that **250 v** would be required at the transformer.

At starting,

$$IX_L = 390 \times \tfrac{600}{1000} \times 0.1296 \text{ or } 30.3 \text{ v},$$

$$IR = 390 \times \tfrac{600}{1000} \times 0.3952 \text{ or } 93 \text{ v},$$

$$IZ = \sqrt{(30.3)^2 + (93)^2} \text{ or } 97.7 \text{ v}.$$

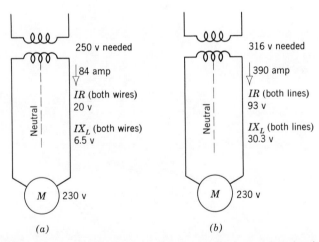

Figure 17-15. Ignoring all other loads on the transformer, we determine its output voltage necessary to maintain 230 v at the motor terminals (a) when the motor is running and (b) when starting. The line is 300 ft long. The transformer voltages required calculate as an impractically high value, indicating that with a fixed transformer output of about 230 to 240 v motor voltage will be very low, especially under starting conditions.

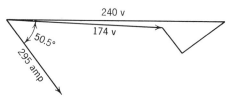

Figure 17-16. If we assume that 240 v is the actual transformer output, the phasor diagram can be reduced in proportion to that of Figure 17-15. Since 240 is 75.8 percent of 316, the drops will reduce in like proportion and the voltage across the motor terminals will be 0.758 × 230, or 174 v. The starting current will then be 295 amp.

Figure 17-17. A large general-purpose, single-phase motor with its end cover removed. By means of this construction the centrifugal switch can be adjusted or replaced or the capacitors can be replaced without tearing down the entire motor installation. No matter how carefully these motors may be built, they represent the "weak links" in single-phase motor construction. (Courtesy of Robbins & Myers, Inc.)

Example   289

Now refer to Figure 17-14 for the phasor diagram. Its solution shows that 316 v would be required at the transformer. This is obviously impossible (see Figure 17-15).

A simple means of determining the voltage across the motor terminals for any transformer supply voltage is by proportion from the diagram of Figure 17-14. Thus, if the actual transformer output is 240 v, the whole phasor diagram can be reduced in the proportion of $\frac{240}{316}$. By this means it can be shown as in Figure 17-16 that only 174 v will be applied to the motor terminals at starting, and the starting current will be reduced to 295 amp.

As the motor is designed for 230 v, the starting torque will now be only 57% of rated torque. This could be too low for satisfactory starting and acceleration (see Figure 17-17).

Chapter 18

An Example of Motor Application and Specifications

This is the case history of the application of an electric motor to a newly developed commercial device. It follows the steps of first samples and production specifications to final production.

The case starts with an inquiry to a motor manufacturer for a single-phase motor of about $\frac{1}{25}$ hp to operate at about 1,700 rpm. The motor will fit into the cabinet of this self-contained machine which is to be belt-driven. The space in the cabinet will accommodate a motor up to 4-in. diameter, and other space is available to mount either a relay or a capacitor. Quiet operation is important, and the environmental conditions are those of the usual office.

18-1 A Tentative Motor Selection

Conferences between personnel of the OEM account and the motor supplier resulted in an agreement on the first trial sample to be submitted. The motor chosen was $3\frac{3}{4}$ in. in diameter because the motor manufacturer had such a frame in production.

For quiet operation the motor would be supplied with a resilient base and have sleeve bearings. A permanent-split-capacitor motor was favored over a shaded-pole motor, because of comparative quietness, and, besides, the shaded-pole motor would lack starting torque for the belt drive.

A split-phase motor was tentatively ruled out because it would not be as quiet as the permanent-split-capacitor type and that much starting torque might not be required.

The cabinet appeared to have poor provision for ventilation. It was expected that a permanent-split-capacitor motor of this size would have an efficiency of about 50%. The shaded-pole motor might be about 30% efficient. As shown in Chapter 5, this would indicate that the first motor would, therefore, have an input of about 60 w of which 30 would be dissipated as heat. The shaded-pole-motor input would be about 100 w of which 70 would help to heat up the motor and the cabinet.

18-2   Motor Specifications

Several samples and some months later, with minor changes made in both the motor and the device, successful operation appeared assured. Prices were agreed upon.

The OEM then asked for a dozen sample motors for further tests. After these were completed, an order was placed for production motors to meet the following specifications:

This motor shall be rated at $\frac{1}{25}$ hp, nominal 1,700 rpm. When operated at 115 v and with a load of 2.0 oz-ft, it shall have a speed of not less than 1,625 rpm. At this load the input watts shall not exceed 58 and the input current shall not exceed 0.63 amp.

Starting torque shall be no less than 1.8 oz-ft. When operating in the machine with 58-w input, the temperature rise measured by thermo-couples on the end coils should not exceed 40°C.

Operating idle in a sound room of suitably low decibel ambient, the sound level shall not exceed 45 db when measured in any direction around the motor with the microphone 8 in. from the center of the motor.

These values were fixed by the OEM engineers testing all 12 of the sample motors and using more or less of an average for each of the items. In collaboration with their quality control personnel, the specifications were then written.

Of the values listed, how many were actually necessary for the successful operation of the machine?

For instance, had the motor been operated at say 105, 100, 95, and 90 v to see if it would still successfully start the machine? Did the motor fail to start at any of these voltages? We have found that torques vary as the square of the voltage. If the motor always started the machine at 100 v but not always at 95 v, this would imply that the minimum starting torque actually required would be:

$$\left(\frac{100}{115}\right)^2 \times 1.8 \text{ oz-ft} = 1.36 \text{ oz-ft.}$$

A compromise might then have been reached between the 1.36 and 1.80 oz-ft which would give the motor manufacturer some leeway in production.

Similarly, was the machine operation really inferior if the speed was less than 1625 rpm? If not, why not consider 1615 or 1610 rpm?

But most important, why limit the load input to 58 w and 0.63 amp? As we shall see, this closed a very important door for future cost savings to the user.

Would the ultimate user really care if the device took 65 or 70 w and perhaps 1 amp?

And why 40°C rise? True, this is an industrial standard for temperature rise, based on 105°C as the final temperature for the insulation. But as a widely sold commercial device this machine will need to be listed by Underwriters' Laboratories. They recognize that the room temperature is usually about 25°C and, hence, permit a temperature rise on the windings of the motor to be 75°C.

Input watts affect the temperature rise; this input need do no harm until it has caused the motor to reach an unsafe temperature. So, if not the full allowance of 75°C rise as permitted, why not a compromise between the 40°C actually observed on the motor and the 75°C top limit?

As has been mentioned before, the motor user apparently believes that a cool motor is a better motor. It really makes no practical difference in the quality or life of an electric motor as to whether it is cool or hot, as long as the operating temperature does not exceed the safe value for the insulating material used, and as long as the hot winding in turn does not heat the bearings beyond the recommended temperature of the lubricant.

And lastly, the sound level. Is the noise really objectionable if the motor decibels are above 45?

18-3 Specifications Set too Soon

Bear in mind that all of these specifications were established by examining what might be called *hand-built samples*. Certain production tooling would be established before large output could be planned and variations in performance would likely occur. Two undesirable procedures have occurred in this case.

Owing to an understandable desire to assure uniform quality in production, the OEM has probably written too thorough a specification. At least some of the values are meaningless except to handicap unnecessarily the supplier.

The other unfortunate happening is that the specifications were written too soon. They should preferably have been written by mutual agreement after several hundred production-line motors had been delivered.

It would seem to be a simple matter for quality control personnel to get together with a supplier and modify or rewrite specifications, examining those items which are really important and compromising on others. But this is not always easy. Once specifications are written, any compromise apparently lays open quality control to criticism that they made a mistake in the first place or that they are now willing to "sell out" for an inferior product.

This results in the continued production of motors meeting unnecessary standards. But before attempting to show how this increases the cost for the buyer, let us examine in more detail what can go wrong with motors on a production line.

18-4   Processing A Motor for Production

Note the stack of laminations of Figure 18-1. These laminations are punched from coiled steel and are then heat-treated to form an insulating

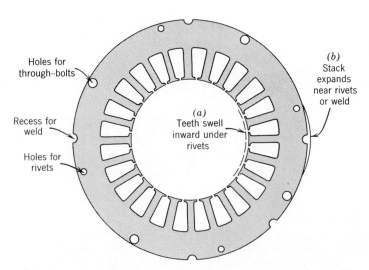

Figure 18-1. A lamination from the die may be nearly perfect in roundness and symmetry. When heat treated to relieve strains and provide an insulating coating, it may become eggshaped by several thousandths of an inch, depending on the size and configuration. When welded or riveted into a stack (usually one or the other, not both), the stack is further distorted by inward or outward swelling, as shown at (a) and (b).

oxide and to relieve the strains that affect the magnetic qualities. In this process a lamination which originally had an inside bore that was round to within $\frac{1}{1000}$ in. may end up egg-shaped by 0.002 or 0.003 in.

The stator laminations are then built into a stack of approximately 1 in. of width, depending on the design. They are held together by welding or riveting. This process usually causes a dimensional change of the laminations radially, so that the outside diameter (OD) and the inside diameter (ID) are now still further from a perfect circle.

The stacks of rotor laminations are put into a mold, and molten aluminum is injected at high pressure and velocity to form the squirrel cage. In this process some of the aluminum cages entrap more air than others, and the result is a slight change in porosity and resistance (see Figure 18-2).

The completed rotor is then ground to the correct OD so as to leave an air gap between stator and rotor as determined by the designer. The rotor is then heat-treated to burn off the aluminum smears on the OD and to break loose the cage from intimate contact with the steel. These corrections minimize stray currents flowing throughout the rotor that can cause negative torques and losses. But the heat treatment can

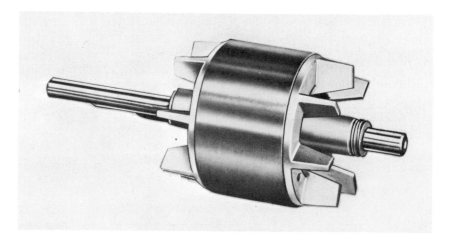

Figure 18-2. Assembled shaft and die-cast rotor. The slots are "buried" below the rotor surface. Rotor laminations are heat treated along with the stator. If they become distorted the slots, hence the die-cast squirrel cage, are closer to the surface in some areas. This encourages pulsation and noise. A small but variable degree of porosity exists in the die-cast aluminum cage. The assembled rotor is turned or ground to the correct OD, heat treated, and pressed onto the shaft. Bearing surfaces on the shaft and rotor OD should be concentric to a high degree of accuracy.

Head fit

Figure 18-3. When the wound stator is pressed into the shell, concentricity between the inside bore of the stator and the ends of the shell must be maintained so that the rotor will be centered in the stator bore. On larger motors the inside section of the shell ends are turned for head fits, registering from the inside bore of the stator.

also distort the rotor so that it is no longer as round as it was after the initial grind.

A check on the shaft and rotor assembly may show that they do not run on the same centers by perhaps $\frac{1}{1000}$ in. and that, furthermore, the rotor is slightly out of round.

Now examine the body or shell into which the wound stator is pressed (see Figure 18-3). The edge of the shell into which the heads are fitted may prove to be eccentric, with the inside bore of 0.001 in.

Registering from the bore of the bearing, fits are turned on the heads as shown in Figure 18-4. Although this is a close-tolerance operation, aging of the iron or aluminum from which the heads are made can distort the head so that the fit is no longer perfectly round nor completely concentric with the bearing bore by 0.001 or 0.002 in. The motor is now ready to assemble, hopefully.

Affected by all of the foregoing ills, it has a slightly out-of-round rotor, running close on one side of its air gap, and is inside a stator

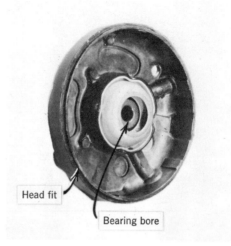

Head fit

Bearing bore

Figure 18-4. Perfect concentricity should be maintained between the turned head fit and the bearing bore. They may be out several thousandths of an inch, depending on motor size. Stress release and aging of the casting can influence concentricity and roundness.

bore which is not round. As compared with a more mechanically true and perfect assembly, it is noisy, has lost some of its expected torque, and may be higher in input current and watts.

The situation described in the foregoing is not caused by careless workmanship, lack of quality control, nor slip-shod processing. It is caused in part by the fact that the very heart of an electric motor—the stator and rotor—is built up of comparatively thin laminations which are subjected to the indignities of heat treatment, welding, riveting, and then all being turned and ground on their edges. They buckle, saucer, belly-out, and deform in protest.

18-5 Variations in Purchased Materials

In addition to the foregoing, the steel used in the laminations can vary greatly in its magnetic qualities. Only the maximum core loss at a given flux density is guaranteed. For instance, it may be 4 w/lb. But the supplier sometimes surpasses himself and ships coils in which the losses may be nearer to 2 w. This is splendid, but it lulls the production line into a false sense of safety. No guarantees are made as to the permeability of the steel. That is, the number of amperes required to bring it up to a certain degree of flux density is variable and not specified.

Magnet wire is, of course, manufactured in various gages and the designer selects the size needed. But magnet wire cannot be produced with an absolutely fixed diameter and some manufacturing tolerance is necessary. It varies from 4 to 5% in cross section (depending upon the size), and this means that the resistance and losses in the winding will vary to the same extent.

The motor under consideration in the beginning of this chapter used a capacitor of, say, 4 μf. When purchased from the supplier, it was allowed a variation of plus or minus 10%. Thus the electrical designer's 4-μf capacitor, upon which he was basing expected performance, might vary from 3.6 to 4.4 μf. This affects the starting and running torque and amperes.

18-6   Precision Motors can be Built

Very precise motors are available, both in the small integral-horse-power sizes, and in the subfractional class of instrument motors. They are precise in their mechanical tolerances and their operating performance. They can hardly be classed as production-line items. Their cost might run from $7.5/lb in the medium sizes to over $100/lb in the smaller sizes. They require no more steel or copper than the normal motor, but their price is determined by close-tolerance workmanship and by an allowance for high rate of rejects, both for the components and the presumably complete motor.

On the other hand, what might be called production-line motors sell for about $1.50/lb in the small sizes, $0.65/lb in medium sizes, and approximately $1.00/lb in the 5-hp 1725-rpm rating.

18-7   Interrelation of Motor Operating Characteristics

We have seen how specifications might be set up by an OEM, and have reviewed the troubles that beset a production line on which these motors are to be built. It is axiomatic in business that the customer must be given what he wants. But, unfortunately, owing either to engineering or quality control, or both, he sometimes receives and finally must pay for what he does not *need*.

Assume high rejection rates occur on the production line and that the main reason is "weak motors"; that is, with a load of 2.0 oz-ft, the speed of 1625 rpm is not maintained as specified.

Furthermore, the starting torque is low on many motors owing to a variety of causes mentioned previously.

The electrical designer, therefore, might decide that more motors would pass production tests satisfactorily if they were made somewhat stronger. This is done, as previously mentioned, by reducing the number of turns in the winding. This reduces the impedance of the winding and increases the current. Current and watts might exceed the limits of 0.63 amp and 58 w respectively, already set in the specifications. This correction may, therefore, be undesirable.

However, as we found in considering the magnetic circuit (Chapter 1), it takes many times the ampere-turns to set up flux in the air gap as compared to steel. The previous high current might, therefore, be reduced if the design were changed so that the rotor OD was increased and the air gap reduced. Unfortunately, a reduction in air gap brings the stator and rotor teeth more closely together. This increases the intensity of the flux pulsations as the rotor teeth move past the stator, and the motor noise increases. An upper limit was set on this.

We have now tried to better the speed and torque by strengthening the motor. This increases the current. To bring the current back to normal, the air gap has been reduced. This increases the noise. Here is a sequence of attempted changes which illustrates the interdependence of one characteristic on another in motor design.

18-8   Who Pays for the Rejects?

With no relief from all inclusive specifications, and the motor designer's hands being tied from making small design changes to improve acceptable productivity, the production line is cursed with a high rate of rejections. Some might be reworked at a high labor cost. Others are completely scrapped. What the buyer does not realize is that, in the long run, he pays for *all* of this.

This comes about because of OEM buyers approach their suppliers periodically for a reduction in price. It is a part of their jobs, even though they know that both steel and copper prices have gone up in the meantime! They reason that the production experience gained by the supplier over, say, a year, must have shown him means of reducing his cost. This is often true. They also have the half-concealed club of bringing competition more heavily into the picture.

The motor salesman, on the other hand, has a fairly accurate figure as to what these motors cost to produce, including the unusual burden of special tests and scrap. He may argue that experience has also shown the customer that the original specifications of some items may no longer be significant and that "easing off" can reduce the production rejection rate and the over-all cost. He may also be aware that a slight saving might be made if lower-cost higher-loss lamination steel were used in the motor. But this would increase the input by perhaps 5 or 10 w and the temperature by a degree or two, and these were all fixed in the original specifications.

He is unwilling to reduce his price below a fair margin of profit, and the two additional means, mentioned previously, by which savings could be made, are apparently unacceptable. His new quotation may represent only a slight price reduction. Thus, in the long run, the buyer pays for the motors scrapped on a production line, most unfortunately so if they are scrapped to meet features of specifications that are *not really necessary.*

Appendix

°C	°F
0	32
5	41
10	50
15	59
20	68
25	77
30	86
35	95
40	104
45	113
50	122
55	131
60	140
65	149
70	158
75	167
80	176
85	185
90	194
95	203
100	212
105	221
110	230
115	239
120	248
130	266
140	284
150	302
160	320
170	338
180	356

Table A-2   Wire Table, Annealed Copper

Wire Size (AWG)	Nominal Diameter (in.)	Cross-Sectional Area		Copper resistance at 20°C (Ω/1000 ft)
		(circular mils)	(in.)	
4/0	0.4600	211,600	0.1662	0.04901
3/0	0.4096	167,800	0.1318	0.06182
2/0	0.3648	133,100	0.1045	0.07793
1/0	0.3249	105,600	0.08291	0.09825
1	0.2893	83,690	0.06573	0.1239
2	0.2576	66,360	0.05212	0.1563
3	0.2294	52,620	0.04133	0.1971
4	0.2043	41,740	0.03278	0.2485
5	0.1819	33,090	0.02599	0.3134
6	0.1620	26,240	0.02061	0.3952
7	0.1443	20,820	0.01635	0.4981
8	0.1285	16,510	0.01297	0.6281
9	0.1144	13,090	0.01028	0.7925
10	0.1019	10,380	0.008155	0.9988
11	0.0907	8,230	0.00646	1.26
12	0.0808	6,530	0.00513	1.59
13	0.0720	5,180	0.00407	2.00
14	0.0641	4,110	0.00323	2.52
15	0.0571	3,260	0.00256	3.18
16	0.0508	2,580	0.00203	4.02
17	0.0453	2,050	0.00161	5.05
18	0.0403	1,620	0.00128	6.39
19	0.0359	1,290	0.00101	8.05
20	0.0320	1,020	0.000804	10.1
21	0.0285	812	0.000638	12.8
22	0.0253	640	0.000503	16.2
23	0.0226	511	0.000401	20.3
24	0.0201	404	0.000317	25.7
25	0.0179	320	0.000252	32.4
26	0.0159	253	0.000199	41.0
27	0.0142	202	0.000158	51.4
28	0.0126	159	0.000125	65.3
29	0.0113	128	0.000100	81.2
30	0.0100	100	0.0000785	104
31	0.0089	79.2	0.0000622	131
32	0.0080	64.0	0.0000503	162
33	0.0071	50.4	0.0000396	206
34	0.0063	39.7	0.0000312	261
35	0.0056	31.4	0.0000246	331
36	0.0050	25.0	0.0000196	415
37	0.0045	20.2	0.0000159	512
38	0.0040	16.0	0.0000126	648
39	0.0035	12.2	0.00000962	847
40	0.0031	9.61	0.00000755	1,080

Table A-3   Ohms Inductive Reactance per 1000 ft, 60 Cycles*

Size, B and S gage	Distance Between Centers (in.)							Amperes†
	6	9	12	18	24	30	36	
10	0.1150	0.1244	0.1309	0.1402	0.1469	0.1520	0.1561	55
8	0.1097	0.1191	0.1256	0.1349	0.1415	0.1467	0.1508	70
6	0.1044	0.1137	0.1203	0.1296	0.1362	0.1413	0.1455	100
4	0.0991	0.1084	0.1150	0.1243	0.1309	0.1360	0.1402	130
2	0.0938	0.1031	0.1097	0.1190	0.1256	0.1307	0.1348	175
1	0.0911	0.1004	0.1070	0.1163	0.1229	0.1280	0.1322	205
1/0	0.0885	0.0978	0.1043	0.1137	0.1203	0.1254	0.1295	235
2/0	0.0858	0.0951	0.1017	0.1110	0.1176	0.1227	0.1209	275
3/0	0.0831	0.0925	0.0990	0.1083	0.1149	0.1201	0.1242	320
4/0	0.0805	0.0898	0.0904	0.1057	0.1123	0.1174	0.1215	370

* For solid conductors. Values for stranded conductors are about 1% less.
† Recommended current capacity for one type of weatherproof, insulated copper wire in free air.

Table A-4   Partial Table of Trigonometric Functions

Degrees	Sine Cosine	Degrees
0	0.000	90
2	0.035	88
4	0.070	86
6	0.105	84
8	0.139	82
10	0.174	80
12	0.208	78
14	0.242	76
16	0.276	74
18	0.309	72
20	0.342	70
22	0.375	68
24	0.407	66
26	0.438	64
28	0.469	62
30	0.500	60
32	0.530	58
34	0.559	56
36	0.588	54
38	0.616	52
40	0.643	50
42	0.669	48
44	0.695	46
46	0.719	44
48	0.743	42
50	0.766	40
52	0.788	38
54	0.809	36
56	0.829	34
58	0.848	32
60	0.866	30
62	0.883	28
64	0.899	26
66	0.914	24
68	0.927	22
70	0.940	20
72	0.951	18
74	0.961	16
76	0.970	14
78	0.978	12
80	0.985	10
82	0.990	8
84	0.995	6
86	0.998	4
88	0.999	2
90	1.000	0

The sine of 20° (0.342) is the same as
the cosine of 70°.

Epilogue

Electrical engineering is more than hardware and the mathematics by which it can be analyzed and synthesized. It also involves people; the people who invented the devices and who struggled and argued about the mathematics that applied to them.

These pioneers are almost forgotten. Responsible as they are in large part for our great industrial complex as well as for the gadgetry of our present civilization, we can look in vain for statues erected in their honor.

Each succeeding generation of electrical engineers seems to know fewer and fewer names connected with its technical inheritance. It might almost be led to believe that its entire technology emerged in full bloom from the mind of its professors as did Minerva from the brain of Jupiter.

This statement is not as extreme as it may appear. Over the course of many years I have hired young engineers who usually appeared at work with an armload of texts and reference books as well as bound copies of notes written by their professors for classroom work. (This is a deplorable practice, needless to say frowned upon by authors and publishers alike.) In the period of a few weeks I usually found time to examine these texts with which I was not always familiar, along with the classroom notes.

The present trend can be epitomized by one example in which I found the usual differential equation for the ac circuit. I asked the young man if he knew any thing about its development and background. He did not, but assumed it was some of the mathematics worked up by his professor for solution of the problems that were presented in the course. He was surprised when I told him that the identical equation and its manipulations were to be found in papers published in the 1890s.

It is not my intention to imply that the writer of such notes had the slightest desire to take credit for that work. In a busy classroom he probably found no time to explain the origin of such fundamental mathematics nor the names of the contributors.

304

A crowded technology has no time left for quoting sources and credits. This has apparently become unpopular and pedantic.

I was fortunate enough to have graduated in the early 1920s. Nearly every engineering faculty had at least one professor who was about 60 years old. That meant that he had graduated in the 1880s. He and his classmates saw at first hand the upsurge in electrical technology that occurred in the ensuing 30 years.

Some of the inventions and developments were accomplished by his classmates or by others whom he knew through technical meetings. They visited the campus for talks on occasion.

And if the old professor was garrulous (and what old professor is not?), his lectures and his conversation would be seasoned at times by remarks such as, "Charlie Scott says this; Steinmetz believes so and so; when I visited Pupin in Connecticut; Lamme first determined that; I saw the policeman bring Oliver Heaviside his dinner and followed him into the entry . . . ," and so on. Not all of the comments were complimentary.

From a modern pragmatic point of view that may all have represented a waste of time. But it was not wasted on me, as it livened everything I have read and worked on in the electrical machinery field for the past 45 years. (It is difficult to tune a radio and not think of Pupin's Serbian shepherds with their daggers stuck into the ground and their ears against the handle.) Only by considerable restraint have the footnotes discussing personalities been held to a minimum in the foregoing text.

As a final note of appreciation for those who included people in their teaching of mathematics and machinery, I wish to thank: Professors Alexander Wurtz and Harry Hower of Carnegie Tech; associates C. A. Wright and F. C. Caldwell of Ohio State; acquaintances Dean E. J. Berg of Union College, Charles Scott of Yale, Thomas Watson who, in assisting Bell, heard the first words spoken over the telephone, V. Karapetoff of Cornell; and my two predecessors in the position from which I recently retired. These were Harvey Stuart, who worked with Edison after graduating from M.I.T. and ex-Professor A. F. Puchstein, who apparently knew every significant professional paper written over a period of 60 years, along with the gist of its contents, who wrote it, and the month and the year that it appeared, as well as the name of the journal. This knowledge extended, but to a lesser extent, over European technology as well.

To all of them my sincere thanks, for making electrical engineering a living subject.

Problems

Chapter 1

Section 1-1

1. A lamp having a resistance of 8 ohms is connected across a 12-v battery. Determine the amount of current that flows through the lamp.
2. A 12½-ohm resistance is connected to a 22½-v battery. What current does the battery supply?
3. A heating element draws 1.25 amp when connected to a 9-v battery. What is the resistance of the element?
4. A 45-v battery is used to operate an electric device. What resistance must the device have in order to draw 1.9 amp from the battery?
5. What size battery is needed in order to make a 2.5-ohm lamp draw 1.8 amp?
6. If a 3.2-ohm load is to draw 7.5 amp, what size battery must be used?
7. A lamp having a resistance of 8 ohms is connected to a 12-v battery using leads each having 1-ohm resistance. What is the current through the lamp?
8. A 12-ohm resistance is in series with a lamp having a 15-ohm resistance. If these are connected to a 22½-v battery, what current flows in the circuit?
9. A lamp having 9-ohm resistance is connected to a 16.5-v battery with wires that each have 1.35-ohm resistance. What voltage appears across the lamp?
10. Two 0.8-ohm wires are used to connect a 2.4-ohm load to a battery. If the voltage across the load is 10 v, what is the battery voltage?
11. A load is operating at 115 v, although the supply voltage is 117.5 v. If the load is drawing 5 amp, what is the resistance of each of the connecting wires?
12. An electric heater is operating from a 220-v supply. If the resistance of the connecting wires totals 0.65 ohms and the voltage across the heater is 216 v, determine the resistance of the heater.

Section 1-2

13. An electrical device draws 1.35 amp from a 120-v source. What is the power supplied by the source?
14. A heater draws 4.5 amp when operated at 9 v. What is its power consumption?

15. One volt is dropped in each wire feeding a 100-w lamp. If the current drawn by the lamp is 0.87 amp, how much power is lost in the connecting wires?

16. One-half volt is dropped in each wire feeding a 75-w lamp. If the lamp draws 2.2 amp from the supply, how much power is dissipated in the wires?

17. What power is dissipated in a 15-ohm resistive element when 0.86 amp flows through it?

18. A 300-w heater draws 2.5 amp from its supply. How much power is dissipated in the wires feeding the heater if they each have 0.52-ohm resistance?

Section 1-3

19. Three resistors—12 ohms, 15 ohms, and 22 ohms—are connected in parallel across an 18-v battery. What is the current drawn by each resistor?

20. A 24-v battery is used to supply four lightbulbs connected in parallel. If the bulbs have 16, 22, 12.5, and 25 ohms of resistance respectively, determine the current drawn by each.

21. A 22½-v battery is used to supply four resistors connected in parallel. If the resistors measure 18 ohms, 22 ohms, 10 ohms, and 15 ohms, determine the total current drawn from the battery.

22. Three light bulbs, measuring 17, 21, and 13 ohms of resistance respectively, are connected in parallel across a 12-v battery. What is the current supplied by the battery?

23. Three resistors, measuring 33, 47, and 39 ohms, are connected in parallel to a battery. Determine their combined resistance.

24. Find the combined resistance of four resistors—68, 82, 27, and 56 ohms—connected in parallel to a battery.

Section 1-8

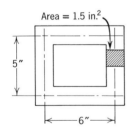

25. A laminated steel core is built up to the dimensions shown. How much current must flow through a 240-turn coil wound on this core in order to produce a flux of 132,000 lines?

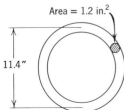

26. A current of 1.53 amp is flowing through a coil wound on a laminated steel core with dimensions as shown. How many turns are necessary to produce a flux of 84,000 lines?

Section 1-9

27. An air gap of 0.02 in. is cut into the laminated core used in Problem 25. What current is necessary in the 240 turn coil to produce a flux of 125,000 lines?

28. An air gap of 0.013 in. is cut into the laminated core used in Problem 26. How many turns are necessary to produce a flux of 120,000 lines if 3.4 amp are flowing through the coil?

Section 1-10

29. A conductor is moved so that it cuts 3.2×10^8 lines of flux per second. What voltage is generated in the conductor?

30. What voltage is generated in a conductor that cuts 560,000 lines of flux per second?

31. What voltage is generated in a conductor that cuts 560×10^6 lines of flux in 2.5 sec?

32. A conductor is moved so that it cuts 5950×10^4 lines of flux every 0.35 sec. What voltage is generated in the conductor?

Section 1-11

33. A coil consisting of seven turns of wire experiences 16.2×10^8 flux linkages in 0.76 sec. What voltage is induced in the coil?

34. What voltage is induced in an eleven-turn coil that experiences 76.5×10^7 flux linkages in 1.72 sec?

35. A conductor is moved so that it cuts 120×10^8 lines of flux every 2.5 sec. If 144 v is induced, into how many turns is the conductor wound?

36. How many turns must a coil have if 197.5 v are to be induced by cutting 1136×10^6 lines of flux in 0.46 sec?

Chapter 2

Section 2-2

1. Consider the line 100 units in length used in Figures 2-2 and 2-3. Determine the height y when the line has rotated counterclockwise through 40°.

2. The 100-unit-long line used in Figures 2-2 and 2-3 has rotated counterclockwise through 62°. What is the vertical projection y?

3. A line 85 units long has rotated counterclockwise through 156°. What is the vertical projection y?

4. A 76-unit-long line has rotated counterclockwise through 295°. Determine the height y.

Section 2-3

5. A two pole (one "pole pair") generator rotates at 3000 rpm. What is the frequency of the generated voltage?

6. What is the frequency of the voltage generated by a two-pole generator rotating at 1800 rpm?

7. At what speed does a four-pole generator rotate to produce a generated frequency of 25 Hz?

8. How many poles must a generator have in order to produce voltage at 400 Hz while rotating at 1200 rpm?

9. How many electrical degrees are there in 360 mechanical degrees for a six-pole machine?

10. How many mechanical degrees are there in 360 electrical degrees for a ten-pole machine?

Section 2-5

11. A sine wave of voltage has a peak value of 184 v. What is the effective voltage?

12. What is the effective voltage of a sine wave that has a peak value of 242 v?

13. What is the peak voltage of a sine wave whose effective voltage is 117.5 v?

14. A sine wave of voltage has an effective value of 220 v. What is its peak value?

15. A sine wave of voltage has a peak value of 212 v. What is the average voltage?

16. What is the average voltage of a sine wave with a peak value of 134 v?

17. What is the peak voltage of a sine wave whose average voltage is 105 v?

18. A sine wave of voltage has an average value of 54.5 v. What is its peak value?

19. What is the average value of voltage of a sine wave whose effective voltage is 120 v?

20. A sine wave of voltage has an effective value of 208 v. What is the average voltage?

21. What is the power developed when a sine wave of 175-v peak produces an average current of 1.2 amp?

22. A sine wave of voltage with an average value of 68 v produces a current of 2.6-amp peak. What is the power developed?

Section 2-6

23. Determine the inductive reactance of a 7-henry coil when used at a frequency of (a) 60 Hz and (b) 50 Hz.

24. A 3-henry coil is to be operated at frequencies varying from 380 to 420 Hz. Determine the minimum and maximum inductive reactance of the coil.

25. A coil has an inductance of 0.5 henry. Determine the current that flows when the coil is connected to a 120-v, 60-Hz source.

26. Determine the current drawn by a 1.6-henry coil connected to a 220-v, 50-Hz power supply.

Section 2-9

27. A force of 7.21 lb to the right and a force of 4.5 lb down are both working at a point. Draw the vector diagram and determine the resultant force.

28. Determine the resultant force when a force of 32.4 lb acting in a northerly direction and a force of 21.7 lb acting in a westerly direction operate together at a point. Draw a vector diagram.

29. Find the horizontal and vertical components of a 135-lb force acting at an angle of 37° with the horizontal axes.

30. A 92-lb force is acting at an angle of 115° with the horizontal axis. Determine its horizontal and vertical components.

31. Two signals of 115 v each are out of phase by 60°. Draw a phasor diagram and determine their phasor sum.

32. A 110-v sine wave is 40° out of phase with a 120-v sine wave. Draw a phasor diagram and determine their phasor sum.

33. Determine the phasor sum of two 60-amp currents that are out of phase by 120°.

34. Determine the phasor sum of three 60-amp currents that are symmetrically displaced by 120°.

Section 2-10

35. Find the impedance of a circuit that has a 16-ohm resistor in series with a coil of 25-ohm inductive reactance.

36. An inductor with 96-ohm inductive reactance is in series with a 143-ohm resistor. Determine the impedance of the combination.

37. What current will flow when a coil having a resistance of 7 ohms and a reactance of 12 ohms is connected to a 120-v, 60-Hz power source?

38. Find the current drawn from a 208-v, 50-Hz supply when a coil having 10-ohm resistance and 32-ohm reactance is connected to it.

39. A 37-ohm inductor and a 72-ohm resistor are connected in series to a 45-v, 400 Hz supply. Determine the voltage (a) across the resistor and (b) across the inductor.

Section 2-11

40. A motor operates at a 0.9 power factor. What is the power furnished by a 120-v, 60-Hz supply if the line current is 2.3 amp?

41. Determine the power drawn by a device that operates at a 0.87 power factor while drawing 1.1 amp from a 208-v, 50-Hz supply.

42. An electric motor draws 500 w from a 220-v, 60-Hz supply. If the current is 2.4 amp, determine the power factor.

43. A single-phase motor requires 1200 w when operating at 12 amp from a 115-v, 60-Hz power source. At what power factor is this motor operating?

44. An electric motor causes the current to lag behind the voltage by 31°. Determine the power factor of the motor.

45. Determine the power factor of a circuit that has the voltage phasor at an angle of 162° and the current phasor at an angle of 124°.

46. An electric circuit operates at a power factor of 0.77. If the resistance is 285 ohms, determine the impedance.

47. A 150-ohm resistor is in a circuit that operates at a power factor of 0.896. Determine the impedance.

48. A coil having an inductance of 0.6 henry and a resistance of 105 ohms is connected to a 120-v, 60-Hz supply. Determine (a) the current, (b) the phase of the current with respect to the applied voltage, (c) the overall power factor, and (d) the power delivered by the supply.

49. A 235-ohm resistor is in series with a 1.3-henry coil which has a resistance of 125 ohms. Find (a) the current, (b) the phase of the current with respect to the applied voltage, (c) the overall power factor, and (d) the power consumed by the coil.

Section 2-12

50. A 12-ohm resistor is in parallel with a coil of 8-ohm reactance. Determine (a) the total current drawn by the combination from a 117.5-v, 60-Hz supply and (b) the combined impedance of the two devices.

51. Find the combined impedance of a parallel circuit consisting of a resistance of 57 ohms and an inductive reactance of 83 ohms.

52. Determine the power factor of a parallel circuit consisting of a 25-ohm resistance and a coil with 34-ohm inductive reactance.

53. A parallel circuit consists of a coil having 150-ohm reactance and a 200-ohm resistor. Find the power factor.

54. The power factor of a parallel circuit consisting of a coil and a 15-ohm resistor is 0.92. If the total current drawn from the supply is 5 amp, determine the inductive reactance.

55. A 0.3-henry coil is connected in parallel with a resistor to a 50-Hz supply. If the total current drawn is 2 amp at a 0.86 power factor, determine the size of the resistor.

Chapter 3

Section 3-2

1. Determine the current that will flow when a 60-v, 25-Hz voltage source is connected to a 2.0-μf capacitor.

2. A 15-μf capacitor is connected to a 12-v, 100-Hz voltage source. Find the resultant current flow.

Section 3-3

3. Find the capacitive reactance of a 1.5-μf capacitor used in a 60-Hz circuit.

4. Determine the capacitive reactance of a 6-μf capacitor when it is used at a frequency of 120 Hz.

5. Determine the capacitance of a capacitor whose reactance is 500 ohms at a frequency of 25 Hz.

6. A capacitor has a reactance of 73 ohms at 120 Hz. What is its capacitance?

7. A capacitor has a capacitive reactance of 20 ohms when used at a frequency of 60 Hz. Determine (a) the current drawn from a 120-v, 60-Hz supply and (b) the current drawn from a 150-v, 120-Hz supply.

8. A 200-v, 1500-Hz signal results in a reactance of 6350 ohms with a certain

capacitor. Determine (a) the current drawn under the given conditions and (b) the current drawn if the supply changes to 170 v and 1200 Hz.

Section 3-4

9. A series circuit consists of a capacitive reactance of 45 ohms and a resistance of 60 ohms. Determine the current drawn from a 100-v, 60-Hz supply.

10. Determine the current that will be drawn from a 208-v, 60-Hz power source by a capacitive reactance of 22 ohms connected in series with 4.58-ohm resistance.

11. A resistance of 2200 ohms is in series with a 1.6-μf capacitor. When these are connected to a 240-v, 50-Hz supply, determine (a) current, (b) power factor, (c) power delivered by the source, and (d) voltage drop across the resistor.

12. A current of 0.07 amp results when a 1500-ohm resistor and a 2.0-μf capacitor are connected in series to a 60-Hz power line. Determine (a) the voltage of the supply, (b) the power factor, (c) the power delivered to the combination, and (d) the voltage drop across the capacitor.

13. What size capacitor must be connected in series with a 47-ohm resistor in order to produce a current of 0.75 amp from a 120-v, 50-Hz supply?

14. How large a capacitor must be used in series with a 680-ohm resistor if the current drawn from a 110-v, 400-Hz supply must be limited to 0.08 amp?

15. At what frequency will the series connection of a 2-μf capacitor and a 390-ohm resistor have an impedance of 885 ohms?

16. A 2300-ohm resistance is in series with a 0.1-μf capacitor. At what frequency will the combination draw 0.0435 amp from a 200-v signal?

Section 3-5

17. A resistance of 400 ohms is connected in series with an inductive reactance of 500 ohms and a capacitive reactance of 200 ohms. Determine the impedance of the circuit.

18. Determine the impedance of the series circuit consisting of 900-ohm capacitive reactance, 1200-ohm resistance, and 400-ohm inductive reactance.

19. Find the current drawn by the series combination of a 1.8-μf capacitor, a 670-ohm resistor, and a 2.4-henry inductor connected to a 120-v, 60-Hz power supply.

20. A series circuit consists of a 0.7-henry inductor, a 0.65-μf capacitor, and a 330-ohm resistor. Determine the current drawn from a 100-v supply at a frequency of 200 Hz.

21. A coil with 15-ohm inductive reactance and 5-ohm resistance is connected in series with a capacitive reactance of 22 ohms and a 24-v power supply. Determine (a) the voltage across the capacitor, (b) the voltage across the coil, and (c) the power dissipated in this circuit.

22. A 100-v power source feeds a capacitor having 43-ohm reactance which is in series with a coil having 56-ohm resistance and 75-ohm reactance. Determine (a) the voltage across the coil, (b) the voltage across the capacitor, and (c) the power dissipated in this circuit.

Section 3-6

23. A 12-henry coil has a resistance of 3 ohms. What size capacitor must be connected in series with the coil to produce resonance at 75 Hz?

24. What size coil must be used with a 1-μf capacitor to produce series resonance at 40 Hz?

25. A 2.53-henry coil and a resistance of 600 ohms are connected in series with a 4-μf capacitor to a 440-v, 50-Hz power supply. Determine (a) the current, (b) the voltage across the capacitor, (c) the voltage across the resistor, (d) the power factor, and (e) the power delivered by the supply.

26. A 200-v, 60-Hz power source is used to supply current to a 100-ohm resistance, a 2.35-henry inductance, and a 3-μf capacitor, all connected in series. Find (a) the magnitude of the current, (b) the power factor, (c) the voltage across the resistor, (d) the voltage across the coil, and (e) the power delivered by the source.

Section 3-7

27. A parallel circuit consists of a 40-ohm resistor, a 20-ohm capacitive reactance, and a 10-ohm inductive reactance, all connected to a 100-v, 60-Hz source. (a) Determine the current through each component, (b) draw the current phasor diagram, and (c) determine the current drawn from the supply.

28. A 150-v, 50-Hz power supply is used to provide current to a capacitor with 25-ohm reactance, a 50-ohm resistor, and a 21.4-ohm inductive reactance, all connected in parallel. (a) Find the current through each component, (b) draw the phasor diagram of the currents, and (c) determine the current provided by the supply.

29. A 5-henry coil has a resistance of 100 ohms. At what frequency will resonance occur if this coil is in series with a 2-μf capacitor?

30. A 0.3-μf capacitor and a 530-mh coil are connected in series with a 10-ohm resistor. At what frequency must a 10-v supply operate in order to produce a current of 1 amp?

Chapter 4

Section 4-5

1. Each winding of a Y-connected three-phase alternator generates 65 v. What will a voltmeter connected between two of the terminals read?

2. Determine the terminal voltage of a three-phase Y-connected alternator whose windings are designed to generate 155 v each.

3. A three-phase Y-connected alternator produces 290 v between two of its terminals. What voltage is generated in each winding?

4. Find the voltage generated in each winding of a three-phase Y-connected alternator that produces a terminal voltage of 136 v.

5. A three-phase Y-connected alternator feeds a Y-connected resistive load of 8 ohms per leg as in Figure 4-8. If the current in one of the lines to the

load is 23 amp, what is the voltage generated in each leg of the alternator?

6. Find the voltage generated in each leg of a three-phase Y-connected alternator if the current in one of the lines to a Y-connected resistive load of 2.5 ohms per leg is 4.8 amp.

7. A three-phase Y-connected alternator feeds a Y-connected resistive load. Find the total power delivered to the load if each line carries 2.4 amp and the voltage between each pair of lines is 150 v.

8. A voltmeter connected to two of the terminals of a three-phase Y-connected alternator indicates 230 v. If this alternator is connected to a Y-connected load having a power factor of 0.86, and 5.2 amp flows in each line, determine the total power delivered to the load.

Section 4-6

9. A three-phase alternator delivers a line voltage of 160 v. If a delta-connected resistive load of 25.8 ohms per leg is connected to the alternator, determine the current in each line.

10. A delta-connected resistive load having 29.4 ohms per leg is connected to a three-phase alternator delivering 435 v from line to line. Determine the current flowing in one line.

11. A delta-connected load has a power factor of 0.87 while drawing a leg current of 12 amp. If the impedance of each leg is 18 ohms, determine the total power delivered by a source.

12. A three-phase alternator is delivering power to a delta-connected load. Each leg of the load draws a current of 9.4 amp at a power factor of 0.77 through its 18.5-ohm impedance. Determine the power delivered by the alternator.

13. A three-phase power distribution system has an overall power factor of 0.9 while carrying 24 amp at 660 v. Determine the power consumption of the load.

14. Determine the power drawn by a three-phase load having a power factor of 0.83 if the line current is 17 amp at 260 v.

15. A three-phase delta-connected alternator delivers 2800 w at a power factor of 0.84 at rated load. If the line current is 8.3 amp find:
    (a) The line voltage.
    (b) The volt-ampere rating of each winding of the alternator.
    (c) The voltage that this alternator would deliver if Y-connected.
    (d) The volt-ampere rating of the alternator when Y-connected.

16. The rated load of a three-phase Y-connected alternator is 1500 w while operating at a 0.92 power factor. If the line voltage is 220 volts, find:
    (a) The line current.
    (b) The volt-ampere rating of each winding of the alternator.
    (c) The line current that the alternator could deliver if delta-connected.
    (d) The volt-ampere rating of the alternator when delta-connected.

Section 4-7

17. A step-up transformer has 200 turns on its 110-v winding. How many turns are there on its 440-v winding?

18. A step-down transformer has 400 turns on its 220-v winding. How many turns must the 208-v secondary winding have?

19. A 4800/240-v transformer operates at 3.2 v per turn. Find the number of turns on each winding.

20. Find the number of turns on each winding of a 440/24-v transformer that operates at 1.6 v per turn.

21. A resistance of 12.5 ohms is connected across the secondary winding of a 120/8-v step-down transformer. What is the current in the primary winding?

22. A 110/230-v step-up transformer has a 3300-ohm resistor connected across its secondary winding. How much current must flow in the primary winding?

23. A 120/440-v step-up transformer draws 12 amp from its supply. What size resistance is connected to the secondary winding?

24. What is the load resistance attached to the secondary of a 220/117-v step-down transformer if 5 amp are drawn from the supply feeding the primary?

Chapter 5

Section 5-5

1. The flux of an induction motor rotates at a synchronous speed of 1200 rpm. If, at no load, the rotor is turning at 1195 rpm, determine:
   (a) The slip speed.
   (b) The percent slip.

2. The no load speed of a polyphase induction motor is 1495 rpm. If the synchronous speed of the flux is 1500 rpm, determine:
   (a) The slip speed.
   (b) The percent slip.

3. A three-phase induction motor operates at a slip of 4% at full load. If the synchronous speed of the rotation flux is 1800 rpm, determine the operating speed of the motor.

4. Find the rated speed of an induction motor that has 3.8% slip when operating at full load. The synchronous speed of the field flux is 1200 rpm.

Section 5-6

5. Determine the synchronous rpm of a four-pole machine that operates at a power line frequency of 25 Hz.

6. A fourteen-pole motor operates from a 220-v, 60-Hz power source. Determine the synchronous rpm.

7. The synchronous rpm of an eight-pole machine is 750 rpm. Determine the frequency of the supply voltage.

8. Determine the frequency of the line feeding a ten-pole machine that has a synchronous rpm of 4800.

9. A two-pole induction motor rotates at 1480 rpm at no-load. Determine the power line frequency.

10. Find the power line frequency that is used with a four-pole induction motor that rotates at 1730 rpm under full load.

11. Find the number of poles for which an induction motor must be wound if its full-load speed is 810 rpm at a power line frequency of 60 Hz.

12. An induction motor has a no-load speed of 480 rpm when operated from a 50-Hz source. For how many poles is the machine wound?

Section 5-7

13. A motor converts 2360 w to mechanical power. What is the horsepower available?

14. What horsepower is produced by a motor that converts 43,300 w of electrical power into mechanical power?

15. What horsepower is developed by a motor that turns at 1750 rpm while producing 25 lb-ft of torque?

16. A motor produces 1.6 lb-ft of torque while turning at 3500 rpm. What horsepower is being developed?

17. A 20-hp, 440-v motor develops 23 hp when rotating at 1710 rpm. What is the developed torque?

18. What is the torque developed by a 5-hp, 220-v motor that is converting 3580 w of electrical power into mechanical power at 1190 rpm?

Section 5-10

19. A three-phase induction motor draws a line current of 13 amp at a power factor of 0.9 lagging from a 220-v line. If the output torque is 14 lb-ft at 1740 rpm, determine the overall efficiency.

20. A three-phase induction motor turns at 1100 rpm while producing a torque of 12.5 lb-ft. Determine the overall efficiency if the line voltage is 208 v, the line current is 12.6 amp, and the power factor is 0.89 lagging.

21. A dynamometer with a 15-in. counterbalanced arm is used to test a $7\frac{1}{2}$-hp, three-phase induction motor. The following data were recorded:
     Dynamometer scale reading, 20 lb.
     Average line current, 12.5 amp.
     Average line voltage, 440 v.
     Total power input, 8200 w.
     Speed, 1700 rpm.
     Power line frequency, 60 Hz.
     Determine:
     (a) The overall efficiency of the motor.
     (b) The power factor at which the motor is operating.
     (c) The number of poles for which the motor is wound.
     (d) The percent slip at which the motor is operating.

22. A dynamometer with a 24-in. counterbalanced arm is used to test a 40-hp, three-phase induction motor. The following data were recorded:
     Dynamometer scale reading, 60 lb.
     Average line current, 47 amp.
     Average line voltage, 440 v.
     Total power input, 30,000 w.

Speed, 1725 rpm.

Power line frequency, 60 Hz.

Determine:

    (a)   The overall efficiency of the motor.

    (b)   The power factor at which the motor is operating.

    (c)   The number of poles for which the motor is wound.

    (d)   The percent slip at which the motor is operating.

23. A 5-hp, three-phase induction motor with a resistance of 2.4 ohms per leg when wired for 220-v operation is tested to determine its losses using the setup of Figure 5-14. A curve such as that of Figure 5-18 is plotted giving the windage losses as 40 w. The following meter indications are recorded:

	No-load	Full-load
Voltage	220 v	220 v
Current	5.9 amp	14.3 amp
Wattmeter 1	−490 w	1600 w
Wattmeter 2	870 w	3155 w
Speed	1796 rpm	1735 rpm

Find:

    (a)   The electrical power delivered to the rotor.

    (b)   The power output.

    (c)   The efficiency.

    (d)   The output torque.

24. Repeat Problem 23 for a motor having a resistance of 1.2 ohms per leg, assuming that all of the other data remain unchanged.

Section 5-16

25. A polyphase induction motor requires a starting current of 65 amp when full rated voltage of 440 v is applied to the terminals. How will a reduced starting voltage of 330 v affect the following?

    (a)   The starting current.

    (b)   The starting torque (express as a percentage of starting torque at 440 v).

26. A three-phase induction motor is designed to run with a delta-connection. In order to reduce the starting current the motor is started with a Y-connection and then switched to delta. Determine the effect of the Y-connection on the following:

    (a)   The starting current.

    (b)   The starting torque (express each answer as a percentage of the delta-connection value).

Section 5-23

27. A polyphase motor has a temperature rise of 40°C while delivering a continuous output of 2 hp. Its losses are 400 w. This motor is used to accelerate and run a 1.5-hp load for 5 min and then coast to a stop. This procedure is repeated 5 times per hour. If the total heat energy developed during each

operating cycle is 200,000 w-sec, for what temperature must the motor insulation be rated?

28. Repeat Problem 27 for a motor that must repeat the procedure 6 times per hour.

Section 5-25

29. The method of "plugging" is used to stop the motor of problem 27. Of the total heat developed during starting and running 30% is due to starting. Determine the temperature to which the motor will now rise.

30. Repeat Problem 29 for the motor of Problem 28.

31. The motor used in Problems 27 and 29 is started, reversed by plugging, and then stopped during a 5 mm period. This is repeated 5 times per hour. Determine the temperature to which the motor will now rise.

32. Repeat Problem 31 for the motor used in Problems 28 and 30.

Chapter 6

1. Explain why the magnetic field produced by the stator of a single-phase motor results in no inherent starting torque.

2. On what basis is the name for a particular single-phase motor derived?

3. A squirrel-cage rotor is placed within the stator of a single-phase motor and power is applied. What will happen if the shaft is turned? What will happen if the shaft is not turned?

4. The Doubly Revolving Field theory of single-phase motor operation is based on what assumption?

5. Why does a single-phase motor not turn initially according to the Doubly Revolving Field theory?

6. According to the Cross-Field theory, two flux systems that are 90° apart in space are set up due to generator action in the rotor. What accounts for the 90° time separation between the two flux systems?

7. Why does a given single-phase motor rotate slower than a similar polyphase motor?

8. Why is the torque produced by a single-phase motor not constant?

Chapter 7

1. What are the names of the windings used in the split-phase motor?

2. Which winding is designed to have a large amount of resistance and low inductive reactance?

3. How are the characteristics of the winding in Problem 2 achieved?

4. Which winding is designed to have a large amount of inductive reactance and low resistance?

5. The voltage applied to each winding is identical in magnitude and in phase.

What then accounts for the difference in magnitude and phase of the respective currents as shown in Figure 7-3?

6. How are the characteristics of the winding in Problem 4 achieved?
7. What happens if the auxiliary winding is not disconnected from the line after it has been used to start the motor?
8. At approximately what speed does the centrifugal switch operate?
9. At what point in the starting sequence does the motor produce the lowest amount of torque?
10. What undesirable effect can a heavy overload have on the operation of a split-phase motor?
11. What type of split-phase motor generally employs a current relay to disconnect the auxiliary windings?
12. Why are current relays not used on all split-phase motors in place of centrifugal switches?
13. What is the ideal angular displacement of the winding flux in split-phase motors? What is the actual displacement in the resistance split-phase type?
14. What is the purpose of the capacitor in the capacitor-start motor? How is it connected to achieve this?
15. Why were capacitor-start motors not generally used until relatively recently even though the fact that they were superior to resistance split-phase motors was well known?
16. How does the starting torque of a capacitor-start motor compare with that of the resistance split-phase motor?
17. How does the torque available during acceleration differ in the two motors?
18. Why must a capacitor-start motor employ a centrifugal switch or current relay?
19. Why do the resistance split-phase and the capacitor-start motors have identical performance when operating at rated speed and load?
20. What compromise must be made when designing a permanent-split-capacitor motor?
21. What advantages are gained by retaining the auxiliary winding and a small capacitance during operation of the motor?
22. What are the advantages of the capacitor-start-capacitor-run motor?
23. What two types of capacitors are used in the capacitor-start–capacitor-run motor?
24. What are the major advantages of the shaded-pole motor? What are the disadvantages?
25. What is the direction of rotation of the shaded-pole motor?
26. Which of the single-phase induction motors do not use a centrifugal switch or a current relay?
27. Which of the single-phase induction motors do not use the split-phase principle to develop starting torque?
28. Which of the single-phase induction motors is the quietest while operating at full load?
29. Which of the single-phase induction motors develops the highest starting torque?

Chapter 9

Section 9-3

1. What is th full-load speed of a three-phase, 18-p, 220-v, 60-Hz synchronous motor.
2. A 110-v, 50-Hz polyphase synchronous motor is wound with 10 poles. What is its full-load speed?
3. What must be the frequency of a 230-v supply to a three-phase synchronous motor with two poles if it is to have a no-load speed of 1500 rpm?
4. A 16-pole, 50-hp polyphase synchronous motor must operate at 450 rpm. What should the frequency of a 440-v supply be in order to achieve this speed?
5. A synchronous motor operates at 750 rpm from a 220-v, 50-Hz supply.
    (a) For how many poles is this motor wound?
    (b) At what speed will this motor rotate when connected to a 220-v, 60-Hz supply?
6. When connected to a 120-v, 60-Hz supply, a 5-hp synchronous motor rotates at 720 rpm.
    (a) At what speed will this motor rotate when connected to a 110-v, 25-Hz supply?
    (b) What is the percent speed regulation of this motor?

Section 9-6

7. A 440-v, 4-p, three-phase, Y-connected synchronous motor rotates at 1800 rpm. At no-load the rotor is 2° behind the stator flux and the voltage generated in each phase is exactly equal to the applied voltage. Determine:
    (a) The rotor lag in electrical degrees.
    (b) The voltage generated per phase.
    (c) The resultant voltage (E_r).
8. Repeat Problem 7 when the rotor lag is 16°.
9. The torque angle of a synchronous motor must not exceed 15° for a certain application. If the full load torque is 120 lb-ft, what is the maximum available torque?
10. The maximum torque available from a synchronous motor that is rated at 95 lb-ft (full-load) is 293 lb-ft. At what torque angle is it operated?

Chapter 10

Section 10-4

1. Find the internally generated voltage of a dc motor when the applied voltage is 120 v, the armature current is 35 amp, and the armature resistance 0.13 ohm.
2. The armature current of a 230-v dc motor is 25 amp. If the armature resistance is 0.28 ohm, determine the counter-emf.

3. Determine the armature resistance of a dc motor that has an internally generated voltage of 112 v while drawing an armature current of 5.5 amp from a 120-v source.

4. The counter-emf of a 27-v dc motor is 23 v. If the armature current is 32 amp, determine the armature resistance.

5. Determine the terminal voltage of a dc generator whose internally generated voltage is 125 v with an armature current of 15 amp and an armature resistance of 0.37 ohm.

6. The armature current of a dc generator is 35 amp. If the generated voltage is 14.8 v and the armature resistance is 0.08 ohm, determine the output voltage.

7. The armature resistance of a 24-v dc generator is 0.0775 ohm. If the internally generated voltage is 28 v, determine the armature current.

8. Determine the armature current of a dc generator that produces 120 v at its terminals with a generated voltage of 128 v and an armature resistance of 0.051 ohm.

Section 10-9

9. A dc shunt motor draws 15 amp from a 120-v supply. The field circuit resistance is 90 ohms and the armature resistance is 0.23 ohm. Find:
    (a) The counter-emf.
    (b) The power input to the motor.
    (c) The power lost in the armature circuit.
    (d) The power lost in the field circuit.
    (e) The electrical power available to be converted into mechanical power.

10. The armature resistance of a dc shunt motor is 0.12 ohm. When connected to a 48-v supply, the line current is 7.6 amp and the field current is 0.64 amp. Find:
    (a) The field circuit resistance.
    (b) The counter-emf.
    (c) The motor power input.
    (d) The electrical power available to be converted into mechanical power.

11. A 120-v shunt motor with an armature resistance of 0.267 ohm has a no-load armature current of 1.5 amp and rotates at 1650 rpm. When a load is added, the armature current increases to 6.75 amp. Determine the new speed assuming that the flux is kept constant.

12. A 550-v shunt motor has an armature resistance of 0.177 ohm and a field circuit resistance of 200 ohms. While rotating at 1200 rpm with no-load, it draws 8.2 amp from the supply. Determine the speed at which the motor will rotate while driving a load that results in a line current of 96 amp.

Section 10-10

13. Using the data of Table 10-1, determine the maximum acceptable no-load speed for a $\frac{1}{3}$-hp, 1725-rpm, shunt-wound dc motor.

14. A $\frac{1}{6}$-hp, 1140-rpm, compound-wound dc motor runs at 1500 rpm with no-load applied. Does this motor meet the standards recommended in Table 10-1?

Section 10-11

15. A 5-hp, 1150-rpm shunt motor has its speed controlled by means of a tapped field resistor as shown in Figure 10-18. With the tap at position 3 determine the speed of the motor and the torque available at the maximum permissible load.

16. A 30-hp, 1750-rpm shunt motor has its speed increased to 3500 rpm by means of a tapped field resistor as shown in Figure 10-18. Determine which tap is being used and what torque is available at the maximum permissible load.

17. The motor of Problem 15 now has its speed controlled by means of a tapped armature resistor as shown in Figure 10-19. If the motor runs at 575 rpm, determine which tap is being used if 18.3 lb-ft of torque are being produced.

18. The motor of Problem 16 is now controlled by means of a tapped armature resistor as in Figure 10-19. If the tap is at position 2 and the speed is 1225 rpm, determine the available torque.

Chapter 11

1. What is a "universal" motor?
2. Why will a shunt motor perform poorly when connected to an ac supply?
3. Why is a shunt motor, with a capacitor connected in series with the shunt field, not qualified to be called a "universal" motor?
4. What is the use of a motor such as that described in Problem 3?
5. What are the differences between a series dc motor and a series universal motor?
6. What precautions must be taken in the selection of brushes for a universal motor?
7. What is the standard direction of rotation of a motor?
8. Why are universal motors chosen for many applications even though it is known that they will not operate on direct current?
9. Why is the high no-load speed of no consequence in most universal motor applications?
10. What advantage does the use of SCR speed control have over series resistance control of speed?

Chapter 12

1. What is the definition of the "ambient temperature"? What is the normal maximum permissible ambient temperature?
2. Why is there an altitude restriction on the operation of motors without derating?
3. What does a "dripproof" machine protect against?
4. How does a "splashproof" machine differ from a "dripproof" one?
5. Why is it important to keep rodents out of electrical equipment?

6. How does a "pipe-ventilated machine" differ from an ordinary "internally ventilated machine"?
7. What do the initials TEFC stand for?
8. What special precautions are taken when using "open" motors in outdoor service?
9. What are the three general categories of ball bearings for motors?
10. What are "slingers"?
11. What is the undesirable effect of excessive side thrust on a sleeve-type bearing?

Chapter 13

1. What do the initials NEMA stand for?
2. Why do most manufacturers observe the NEMA Standards?
3. What advantages do standard "frame" sizes for motors give?
4. How do the frame numbers differentiate between fractional and integral horsepower sizes?
5. How does one determine the distance from the base to the center of the shaft from the frame number (a) for a fractional horsepower machine and (b) for an integral horsepower machine?
6. In single-phase motors, what determines the horsepower rating?
7. What does the service factor tell about a motor?
8. What does the code letter on a motor indicate?
9. How does the motor "code" letter differ from the "design" letter?
10. According to Table 13-5, which ¾-hp motor would have higher starting current: (a) a 220-v, code B motor or (b) a 440-v code J motor?
11. What is the basis for the Underwriters Laboratory tests?
12. What does a "yellow card" listing by UL mean?
13. How do explosionproof motors achieve this aim?

Chapter 14

1. What is the source of the heat that must be dissipated from a motor? Why must it be dissipated?
2. What determines the "life" of an electric motor? How can the life be prolonged?
3. What determines the allowable temperature rise of a motor winding?
4. How may the winding temperature be determined if thermometers are not available?
5. How do the temperature rises measured by thermometers compare with those measured by the method of Problem 4?
6. How can "burnouts" due to blocked ventilation be prevented?
7. When is an "automatic reset" thermal protector undesirable? When is its use acceptable or even desirable?

Chapter 15

1. How does a "definite purpose motor" differ from a "special purpose" motor?
2. How does a "definite purpose motor" differ from a "general purpose" motor?
3. Give three examples of "definite purpose" motors.
4. How are hermetic motors cooled? What disadvantage results from this? What advantage?
5. Why are the washing-machine-type motors the least expensive for their size?
6. Give three examples of "special purpose" motors.

Chapter 16

1. What is the outstanding characteristic of servomotors?
2. How is the characteristic achieved?
3. What is the ideal torque characteristic of such motors? When does it occur in practice?
4. Where are instrument motors used?
5. What determines whether a servomotor or an instrument motor is to be used?

Chapter 17

1. Why did the oil-pumping industry require large single-phase motors, whereas the other industries did not?
2. What is the factor that becomes a problem with large single-phase motors?
3. How is a "soft-start" generally accomplished when reduced voltage is not used?
4. Why are magnetic contactors used rather than centrifugal switches?
5. What is the main difficulty with the soft-starting of motors? When is this no problem?
6. What is the advantage of the repulsion motor over the capacitor-start motors?
7. How is the direction of rotation of a repulsion motor changed?
8. What unique component does the repulsion-start induction-motor have?
9. How does the repulsion-induction motor differ in (a) construction and (b) operation from the repulsion-start induction-motor?
10. What are the main disadvantages of the repulsion-type motors?
11. The development of what device did generally lead to the phasing out of the repulsion-type motor?
12. What function does a "static phase converter" perform?
13. What is a wire size that meets National Electric Code requirements not always acceptable in practice?

Answers to Odd-Numbered Problems

Chapter 1

1. 1.5 amp
3. 7.2 ohms
5. 4.5 v
7. 1.2 amp
9. 12.7 v
11. 0.25 ohm
13. 162 w
15. 1.74 w
17. 11.1 w
19. 1.5 amp; 1.2 amp; 0.818 amp
21. 6.023 amp
23. 12.96 ohms
25. 1.376 amp
27. 3.09 amp
29. 3.2 v
31. 2.24 v
33. 149.1 v
35. 3 turns

Chapter 2

1. 64.3 units
3. 34.6 units
5. 50 Hz
7. 750 rpm
9. 1080°
11. 130 v
13. 166.2 v
15. 135 v
17. 165.1 v
19. 108 v
21. 165 w

23. 2640 ohms; 2200 ohms
25. 0.636 amp
27. 8.5 lb
29. Horizontal = 107.9 lb; vertical = 81.2 lb
31. 199 v
33. 60 amp
35. 29.7 ohms
37. 8.63 amp
39. 40.1 v; 20.55 v
41. 199 w
43. 0.869
45. 0.788
47. 167.5 ohms
49. 0.395 amp; −53.7°; 0.592; 19.5 w
51. 2.13 amp; 46.9 ohms
53. 0.6
55. 55.8 ohms

Chapter 3

1. 0.01885 amp
3. 1770 ohms
5. 12.74 μf
7. 6 amp; 15 amp
9. 1.336 amp
11. 0.0809 amp; 14.4 w; 178 v
13. 20.8 μf
15. 100 Hz
17. 500 ohms
19. 0.1365 amp
21. 61.4 v; 44.1 v; 38.9 v
23. 0.469 μf
25. 0.783 amp; 623 v; 440 v; 1.0; 344 w
27. 2.5 amp; 5 amp; 10 amp; 5.58 amp
29. 5040 Hz

Chapter 4

1. 112.5 v
3. 167.5 v
5. 184 v
7. 623 w
9. 10.74 amp
11. 6760 w
13. 14,250 w

15. 232 v; 1112 v-amp; 402 v; 3340 v-amp
17. 800 turns
19. 1500 turns; 75 turns
21. 0.0427 amp
23. 134.7 ohms

Chapter 5

1. 5 rpm; 0.417%
3. 1728 rpm
5. 750 rpm
7. 50 Hz
9. 25 Hz
11. 8 poles
13. 3.16 hp
15. 8.32 hp
17. 70.6 lb-ft
19. 0.777
21. 0.735; 0.872; 4 poles; 5.56%
23. 3196 w; 3040.5 w; 0.64; 12.36 lb-ft
25. 48.75 amp; 56.2%
27. 27.8°C
29. 52.3°C
31. 61.1°C

Chapter 9

1. 400 rpm
3. 25 Hz
5. 8 poles; 900 rpm
7. 4°; 254 v; 20.35 v
9. 464 lb-ft

Chapter 10

1. 115.45 v
3. 1.43 ohms
5. 119.45 v
7. 51.6 amp
9. 116.86 v; 1800 w; 42.9 w; 160 w; 1597.1 w
11. 1632 rpm
13. 1983.5 rpm
15. 2930 rpm; 9.14 lb-ft
17. tap 2

Index